Knife Wounds From Desert Sand

Knife Wounds From Desert Sand

Dark Wars Behind Contractor Support in Iraq

BY

JOHN CHANCEY KINGSTON

Published in the United States of America by
Heritage Prime Investments, Inc.
San Antonio, Texas

Knife Wounds From Desert Sand:
Dark Wars Behind Contractor Support in Iraq

Cover Illustration & Interior Design by:
Your Choice Creative Design
yourchoice.creativedesign@gmail.com

ISBN
Hardcover 978-0-9906113-0-1
Softcover 978-0-9906113-1-8
Ebook 978-0-9906113-2-5

PART ONE

GETTING TO THE WAR ZONE

CONTENTS

To all of the fallen military and civilian contractors who served honorably in both Operation Iraqi Freedom and Operation New Dawn

PREFACE

In May 2009 the Iraq War was in its sixth year with no clear end in sight. The conflict began with an invasion under President George W. Bush's Administration. This war was based on the President's discernment of Iraq's possession of Weapons of Mass Destruction. In spite of an extensive search throughout the country, to date, no weapons were ever found. The responsibility of the war now resided with President Barack H. Obama and the new President's position of the war would be critical as to when and how the conflict designated as *"Operation Iraqi Freedom"* would end.

ACKNOWLEDGEMENTS

`This book is based on true events and was written in real time. Several emails have been included to authenticate positions taken within this three part writing. The name of the actual corporation words of employment is not used. The actual name of the Army Program is not used. The actual name of the Military Task Force is not used. The tangible names of the contractors on this program are also not used; any similarities are purely coincidental. However, all sites, bases, cities and countries are never substituted. In order to honor two soldiers, their name will remain unchanged. The words used to illustrate this story were initially inspired by the observation of a leader who appeared to be Baghdad's worst in the WarWizard Program. However, as the observation period continued, other leaders and coworkers emerged far worse than the original which sometimes left me wondering how a number of them were able to be a part of such an important program. I soon learned that many could not be trusted, but I had to be silent with what was discovered because there were enemies on our team and I could not yet distinguish friend from foe; causing the birth of this manuscript.

The pages to follow will portray some of my moral convictions with a first person account of honor, duty, leadership, management, competency and betrayal based on my background as a retired United States Air Force, Master Sergeant now serving as a United States Army Contractor; as observed in both Iraq and Kuwait.

I must thank the many people who were an indispensable part of this books development and theme.

First I must thank my wife, Gwendolyn, for her very professional suggestions, thoughts, creativeness and revisions.

I must also thank Mrs Britney Laroche for her professional suggestions and extremely systematic but thorough edits.

Finally, I must thank former Air Force supervisor and good friend CMSgt (Ret.) Larry E. Harper for our many military discussions and observations of the Peter Principle at work in our unit which helped set the tone and mindset for this writing.

INTRODUCTION

When the average American hears of a fallen soldier, usually their perception is that the fallen member probably died while firing their weapon in a heated firefight on the battlefield. In some cases this does indeed occur, but not as often as one might think because the battlefield has drastically changed; much in part because America posses a large and powerful military; A straight-on fight would be suicide for most opposing forces. Therefore, after the Iraqi military was quickly decimated, the remaining Iraqi opposition used an old but effective resistance when facing such odds; Guerilla Warfare Tactics (roadside bombs, sniper fire, scud rocket attacks, sabotage etc.). The effects of Guerilla Warfare matters were described forty years earlier when Henry Kissinger was quoted to say *"The guerrilla wins if he does not lose. The conventional army loses if it does not win."* Subsequently, many cases of injury and the loss of American life occurred due to either a soldier stepping on a prepositioned landmine or a road side bomb detonation against our convoys.

To resist the new explosive threat, a Combined Joint Task Force named Thorn was established from elements of the Army, Navy and Air Force. Thorn's *Explosive Disposal Forces* specialized in Counter-*Improvised Explosive Device* (IED) measures to eliminate explosive hazards. But the Counter-IED portion of the program that protected military convoys was called WarWizard.

The WarWizard Program was one of the critical United States Army Programs that carried out this task by providing Electronic Frequency Jamming Protection for U.S. military convoy vehicles against roadside bombs. American corporations provide the manpower and skill through the use of the civilian contractor; and that's where I got involved. We were a unit of civilian contractors. A military commander was in charge, but the technicians were all civilian contractors and government employees. Our unit's mission was to prevent as many of those roadside bombs from exploding as possible. We performed this electronic countermeasure task by installing Electronic Jamming Systems in the military convoy vehicles.

The Contractor played a very unique but significant role in the war and just as the military many civilians lost their lives. Considering the mission that had been assigned to the military and the contractor, one would think that we would have our brightest and most dedicated leaders to conduct such an important mission. Often that was clearly not the case. We all innately possess the human ability to make mistakes, but with the consequences of the task at hand, we couldn't afford to make mistakes in this area because American and ally lives directly depended on us.

I have encountered many accomplished contractor managers and leaders within the program. However, to my surprise I have observed a level of sinister behavior that seemed to be replicated more within this program than anyone could deem imaginable. Subsequently, this program appeared to reward the most corrupt by often selecting them for advancement.

I will take you back to the war zone where you can watch the unveiling of "Knife Wounds From Desert Sand" and how the wounds interfaced with key managers who sometimes have long surpassed any honor or loyalty at any level.

CHAPTER 1

LIFE REDIRECTED

A female co-worker and I were selected and given a plane ticket bound for another Middle Eastern country. Each of us carried a thick brown packet to hand deliver. The packets were sealed when given to us and we were instructed for the packet to remain sealed when it was delivered to a man in a red cap after we landed. But when it was time to board the plane I hesitated to board. In this environment, I was not sure if I could trust the packets content.

What if a mistake had been made with my packet? What if there was something other than what I was told in the envelope? I am sure that the authorities at the airport would not believe that I was not aware of the contents if something illegal was being transported. What would I say to my family if the next time I spoke to them was from a Middle Eastern jail or even worse, prison? I then began to reflect on the initial events that transpired that brought me to this time and place...thus my story begins.

May 2009 proved to be a life changing experience when I was hired by a large organization for a position in Iraq. This leg of the company (electronics/ communications coupled with maintenance) paid a high six figure salary. After a year of "pounding the pavement" looking for a job with no offers on the table and the current recession well on its way, the timing of the position could not have been better. My good friend Bert had been working for Communication Systems Services (CSS) for over three years and was the person who linked me with the company. During this time there was a Swine Flu Pandemic occurring. I lived in Texas, which was the state where the first two cases of the disease appeared in the United States.

After a background check to obtain a security clearance, a physical, drug screening and a plane ticket, I was on my way. A few hours later I arrived at the Washington Dulles International Airport. A rental car had been reserved for me by CSS. After getting possession of the car I headed out for my forty mile drive to Falls Church Virginia for their new personnel hire processing and orientation. It was about 10:30pm when I reached the Marriott Hotel which was also arranged by my new employer. Exhausted from the trip and anxious about the days ahead, I set my alarm, took a long hot shower and within minutes I was asleep for the night.

The next morning my phone rang at 6:45am. The caller uttered "Is this John Kingston?" "Yes" I responded. "Are you a new hire for the CSS organization" he questioned? I again responded "yes." "My name is Bryan Bass and I am also a new CSS Employee. Do you mind if I catch a ride with you?" "No, I don't mind at all, what will you be wearing and how will I recognize you, Are you Black, White, Mexican or Asian" I questioned? The voice on the phone replied "I am White about 5 feet 8 inches and will be wearing a blue checkered shirt, blue jeans." I responded, "We are about the same height and I will be wearing brown slacks and a tan shirt." We ended the phone conversation by agreeing to meet down stairs in fifteen minutes. Meeting Bryan proved to be a pleasant experience; it was as if I had known this guy for years. Bryan was about seventeen years younger than me, but we clicked right off the bat. Bryan relayed to me that he was a former Army Infantry Soldier, and had worked several other contracts in the electronic field.

After a few minutes of introductions and dialogue, we set out to find our appointed destination. We drove around in what was no more than a quarter mile radius only to discover there were many buildings bearing the name of our new company. We quickly realized this was not going to be a smooth order of events. The company gave us no concise instructions as to where to report.

After several stops and several buildings later, we finally made our way to the correct building for in-processing. This in-processing took about four hours. During our time there, we were told that if the Program

Director, Edmond Gibson, was available we would be able to meet him. I looked forward to the possibility of meeting Mr. Gibson because on the two occasions I had spoken to him on the phone, he was quite professional. Therefore, I was a bit curious as to whom the man was behind the voice. As the day progressed, I met with Olivia Mitchell. She was one of the hiring managers that I had been corresponding with on getting my package facilitated prior to this point. She was an African American woman around age fifty. She wore her hair in braids and was nicely formally dressed. She was slightly heavy but wore it well on here 5 foot 7 inch frame. I have always viewed Olivia as polite, knowledgeable and professional and today there was no exception. Her positive and professional attitude was very much present as she informed us she would do her best to get us a chance to meet Mr. Gibson, but warned us he was quite busy today. Finally, forty minutes later Olivia escorted us to an area of the building and told us Mr. Gibson would meet with us within the next five minutes.

"Hello and welcome to Communication Systems Services, I am Edmond Gibson the Program Manager" the big statue of a man exclaimed. I had spoken with Mr. Gibson two times on the phone and based on his voice I mistakenly thought he was white; but to my surprise he was an African American. "You men just come on in and I would like to hear a little about each of you and after I have heard from each of you I will tell you about me and moreover about this organization." Earlier that day during our *in processing*, Bryan and I met Willie Venture (a thirty-two year old African American male and former Marine with a background in Electronics and Radio). The three of us, Bryan, myself, and Willie, filed into Mr. Gibson's office and began unfolding small pieces about ourselves. We all chattered enthusiastically about ourselves for a moment then relinquished the conversation back to the approximately 6 foot 3 inch, 260 pound man.

Mr. Gibson began to speak "Well gentlemen I am a family man and have been married for thirty-three years. If you look around the room you will notice there are three pictures. I like to call these three the three phases of upward movement for African Americans. In the first picture you see us just after slavery (a picture of an African American farmer). During this time we were farmers because for the most part that was all we knew. The next phase was the introduction of the fighting man (a picture of an African American soldier). As you know it was around 1860 when the African American soldier was finally allowed into the military. He was then able to get into the fight for his rights. The final phase is where we are today (a picture of a black figure in a high position reaching down to pull up another black figure). We are finally in a position to truly help each other and it is up to each of us to emulate this message".

Mr. Gibson continued speaking by introducing us to CSS. "As far as the company, it is a Fortune 500 company that has been in existence since the 1960s. The project you are joining is just one leg of many legs that extend throughout the world. It is a good time to join because we can always use good leaders in the field. All three of you are headed to Iraq. Now from my prospective, we CSS may have a strong presence in Iraq for about two more years. This is all based on what President Obama says he is going to do with his withdrawal plan. From there, the fight shifts to Afghanistan. None of this is firm of course; we are making these assumptions based on the message coming from the White House.

Now as far as our firm, we have a great healthcare and dental program. A matching funds 401 plan, and if you stay with us for three years you are vested. So as I stated it is a good time to join our team. Do you have any questions for me?" "Yes" Willie replied, as he stood. What rank were you when you were in the military?" "I was a Colonel" replied Mr. Gibson. Mr. Gibson closed the meeting by again welcoming us to CSS. We in turn thanked him for his time. I walked away with a great deal of enthusiasm about my future with this organization.

Our next feat of that day was to check out of the Hotel and catch our flights to our next destination. We all traveled to the Airport together. Upon arrival to the airport, Bryan and I immediately checked our luggage so that we would lighten our load while we enjoyed a leisurely meal. We stopped in at a Five Guys Hamburger Restaurant and found their burgers to be really good.

We then proceeded to the gate section to prepare for departure, however, the flight for Bryan and I had been canceled. Willie's flight had been scheduled on a different airline and was still on schedule. "This sort of thing always happens to me" Bryan voiced. "What are you talking about?" I questioned. "Stuff like flights being canceled" Bryan continued. "Well Bryan, if both our flights got canceled and this almost never happens to me, it's not just you," I responded. After we discovered the flight cancellation we immediately began inquiring about the next flight out for New Jersey. The answer was not one that we were prepared to hear. The clerk informed us that there would not be another flight until the following day at 4:00pm. We both then questioned "What about our luggage we've already checked?" "Your luggage should be there about 11:00pm tonight" responded the clerk. We had to be in class at 8:00am the following day. If we did not receive this initial training, we would have to wait until the next scheduled class date which was unknown to us. Bryan and I knew missing this class was not an option since we both had already experienced about twelve months worth of delays prior to getting to this point. Therefore, after deliberation Bryan and I decided to rent a car and drive to New Jersey to be in attendance for tomorrow mornings training. "This always happens to me" Bryan stated again. "What do you mean?" I questioned. "Flights canceled, luggage missing, next there probably won't be a rental car available for us; I just have bad luck" Bryan complained repeatedly. "Well I usually have good luck and that same plane canceled on me as well" I voiced. "I just rubbed off on you" Bryan replied.

Getting a rental vehicle without a reservation was no easy task because each agency appeared to be booked solid. Bryan and I were to the point of total frustration. Hertz, Alamo, Avis the list went on until we had checked nine different companies. "I told you I am just bad luck" Bryan uttered. "Man stop saying that, luck is what you make it, and we are not done yet" I voiced. As fate would have it, the very last company that we tried also did not have any automobiles available, but had an agent who went above and beyond what was required. He was a very dark African American Man about sixty years old, with a Jamaican accent. He did not have a car for us to rent, but he performed some sort of internal check with all the companies via phone and after going through about five additional rental companies, the sixth one had a car. Bryan and I were overjoyed and thanked the gentleman from *Budget Car Rental* for helping us out of a huge jam. I tried to tip him but he would not take it. I then boasted "You see Bryan, Your luck has already changed."

CHAPTER 2

AN UNPLANNED DRIVE

I don't remember the exact model of the car that we rented other than it was a Chevrolet and it was small; in fact it was very small. It was so small, Bryan's oversized luggage bag was the only thing we could fit in the trunk. However, the size of the car was a non-issue for us, we were just glad to have a car, so I simply placed my luggage on the back seat. Besides this was the most fabulous auto around as far as the two of us were concerned because it was the only car available. Being that we were about to embark on unfamiliar territory, we attempted to obtain a navigation system along with the auto, but of course they were all rented. At any rate, based on the distance that we had to go we estimated that it should take about five hours to make the journey. I hopped in the driver seat and Bryan on the passenger side, and away we went. It was about 600pm and early in our drive we realized we had driven about seven miles in the wrong direction. Back in Texas this would not have been that big of a deal, but we had to drive about fifteen additional miles before we could find a place to turn around and get back on track.

During our drive, Bryan and I continued to exchange excerpts from our lives. Bryan and I both lived in the same region of Texas but with about ninety miles separating our cities. Bryan is married, but admitted he and his wife were not on the best of terms. Of course my next comment was "What is wrong?" In his words he described his wife as black and her family really did not completely accept him. Bryan continued, "The one who rejects me the most is her brother and the really odd thing about that guy is he is married to a white girl." Bryan further explained, "I am ok with her mother though, her mother is very sweet. She is always trying to feed me. On several occasions I had to sit and eat Greens her mother offered to me because I did not want to offend her. Her dad isn't' too bad but he drinks a lot. Her brother turns very nice when he wants to borrow money." I interrupted, "Do you and your wife have any children?" "No we do not have any and that may be the problem" Bryan replied. "Yes if your wife wants kids that may be the reason. How long have you guys been married?" I questioned. "Five years, sometimes I just don't know what to think" Bryan answered.

Bryan continued to unfold "Sometimes I think my family is just as bad." "My sister lives in Florida with my mom but does not give her any money for staying there. She continuously takes what she can from my mom. I am not saying that she steals or anything like that, but she just does not contribute anything. My mom is very sick and I guess you can say that my sister takes care of her. But the problem I constantly see is whenever I come to town the house is always messy. Usually a good hot meal is not being served to my mom; just some quickly prepared junk. Also, if my mom goes into the hospital my sister usually does not stay with her; instead, she is usually hanging out with her friends or doing something unproductive. Most of the time me and my sister don't talk or have little to say to each other." "Man, that's not good. Do you have any other brothers or sisters? I questioned. "No, it is just me and my sister and my mom. My sister does have a daughter, so I am an uncle" he voiced.

"What about your dad, is he still alive?" I again questioned. "Yeah he is alive, but I do not have a relationship with him. He is a loser. He abandoned us when I was about three years old. My sister was only about six months at the time. He showed up after being gone for about twelve years. He shows up every now and then when he wants some money, but other than that, we don't hear from him. He and my sister try to have a relationship, but I want nothing to do with him. Bryan explained.

Bryan further stated "The house that my mother lives in was paid off by me and I have the title to the property. I pay all taxes and when the utilities get behind, I pay them, not my sister. My mother gave me *Power of Attorney* of her estate. My sister wants me to hand it over to her because she lives in town, but my mom assigned me as her Power of Attorney for one simple reason, she knows that I always have her best interest in mind. My sister will sell her stuff without her permission. When I see this occurring, I stop it. But I don't live in town, so I know this continues when I am away. To give my sister Power of Attorney of my mom's estate would be a disaster."

Again I interrupted Bryan, "Man Bryan, I am sorry to hear that. After hearing all those hardships within your family, I guess my blessings I've taken for granted need to be really appreciated. I come from a family of five in which I am the oldest. I have four brothers and one sister; and even today she is still a little spoiled. But she is the only girl so it's understandable. I continued our concentrated conversation by sharing more of my family dynamics with Bryan. "We all get along with each other, but as with any family you have your favorites. My youngest brother is eight years my junior. Though we all are very similar, of the boys, I seem to have the most in common with my youngest brother which hastens a more vigorous relation; but my sister and I have generally the same vantage point and thought process on most issues which always makes a solid relation. I told you that I was married; well I have been happily married for twenty-eight years. I have two kids; a daughter who is twenty-seven and a son who is twenty-five. My daughter is married and has afforded me and my wife with our three Grandchildren; all girls."

At that instance I remembered something that my childhood best friend's dad once said to me over thirty years ago when I was a year into my marriage and military service. The conversation back then went along these lines. " Chancey, I am willing to bet that as long as you associate with married men with families you probably will be able to relate to them more than unmarried guys" Mr. Morse voiced. "I then responded, I can get along with just about anyone but my wife and I socialize best with other married couples." Even though Bryan was seventeen years my junior we conversed as if we had known each other for years.

We finally made it to Maryland. Two miles into the state, we saw our first car pulled over by the highway patrol. One half mile further another car was pulled over; then another and another. We quickly realized it would be best to ensure that we were not caught up in that bunch of ticketed vehicles therefore, I slowed down. As I continued to drive, we both noticed there were stretches of highway twenty-five to thirty-five miles or so without road signs to identify the highway. In addition to the lack of road signs these were some of the darkest stretches of interstate I had ever experienced.

As we continued on our journey the conversations continued to flow. Bryan mentioned that he wanted to see the new Transformers movie sequel that was being released that week. He then suggested we should check it out together if I didn't mind. I responded by saying, "No I don't mind going. I'm not much of a Transformers fan but I did like the first one." Bryan questioned me on the type of movies that I do like. I responded, "Sci-Fi, Documentary and Historical such as *the Last of the Mohicans* or Classics like *Inherit the Wind* are the type of movies that interest me. However, when I just want to relax I will sometimes watch something comical or movies with a humorous theme like *Oh Brother where Art Thou?* With the Soggy Bottom Boys, or *Bad Santa, My Cousin Vinnie,* and who can forget *Bobby Boucher* in *The Water Boy*. So the transformers movie will be O.K. with me." "Great!!!!!" Bryan responded with excitement, "I really would like to see the new one, it starts this Wednesday. I will look into it as soon as we get settled in tomorrow."

As we continued our intense darkness drive, we realized there were very few cars on the highway. We had been on the road for about three and a half hours and were noticing interstate signs were still few and far between. I really began to wonder if we were on the right road. Bryan was not as talkative as he had been. I am sure his lack of words was because he as I were not sure if we were still on course. Had we taken a bad turn? Would we be able to start the class the next morning? Things really looked bleak. We were going around a blind curve and suddenly a sign appeared as if someone had just stuck it there. The sign directed us to turn right *quickly*! I almost missed that turn, but luckily I turned immediately and made the cutoff onto the new highway. The new highway took us over a bridge, and at the top of the bridge was a large illuminated sign with the words

New Jersey and an arrow pointing to the right. We both uttered a sigh of relief after the sign confirmation that our wheels were spinning in the right direction.

It was about 11:30pm when we started seeing commercial aircraft in the landing pattern of an airport. As we got closer we could see that this was clearly the outer portion of the aircraft landing pattern at the New Jersey Airport. Finally, we made it to the New Jersey airport to return the rental car we were driving and then picked up the rental car previously reserved for use by our new company during our training stay. Our next task was to gather our luggage. Bryan started to doubt his luck again by saying, "I bet my luggage was lost" "I told you your luck was changing" I reassured Bryan. According to the airline attendant back in Washington Dulles, our luggage should have beaten us by about three hours. Bryan quickly found his bags; however, my bags were not present. We looked everywhere but my bags were not found. "See, the bad luck has rubbed off me and on to you" Bryan insisted. "You know you need to stop that" I uttered quietly. But, in the back of my mind I was thinking, "I have flown for years and never have I had luggage lost, and as soon as I meet mister bad luck my bags are missing, Hummmmmmmm." I filed a claim with the airlines and could only hope I would soon be reunited with my belongings. We headed to our reserved rooms for a well deserved hot shower and rest at the Holiday Inn.

CHAPTER 3

NEW JERSEY

The next morning I noticed a considerable difference in the temperature in New Jersey than the temperature I left behind in Texas. It was mid June and the temperature was about 94 degrees when I left home. New Jersey's temperature was about 74 degrees; a little cool for June I thought. Nevertheless, it was a great morning. So Bryan and I ventured off to our training that was located on Fort Monmouths Army Post. After all of our encounters the previous day, we both were thankful to be in attendance. The course was designed to give basic knowledge of the systems that we were going to install and maintain. When we got to the classroom there was Willie Venture sitting at the table. The class consisted of a total of ten students. Willie, Bryan and I were employed by CSS. Mark was a sixty year old Caucasian gentleman from Florida who was also a CSS employee. There was a girl and guy from EAB (another oversea contacting house) and four other students (three guys and a girl) from Tommy Hagar Army Depo. "Hi, I am Gabe" voiced the instructor. Gabe was a retired Marine. After Gabe told the class the year in which he retired, doing some simple math let us know that Gabe was about sixty-nine years old. He stood about five feet nine inches tall and was about 175 pounds. His physical stature was that of a man of about thirty-five. Gabe had a training class assistant instructor, also a retired Marine.

After everyone was introduced, the instructors gave an overview of the course and course instructions were underway. Our CSS group had grown to four with the addition of Mark. When we broke for lunch we stayed together and ordered lunch from an Italian Restaurant. We spent this time exchanging stories while basking in what I viewed as perfect weather. The temperature was holding at about 75 degrees at noon. Back in Texas it was about 20 degrees hotter. The people, the temperature and the greenery of the area really made this place somewhere I could get use to. At the end of the day our newly formed CSS group relaxed at dinner and later exchanged reviews and notes from that day's lesson.

The next day was strictly hands-on training on the different equipment. We got out of class about an hour early. So instead of going straight to dinner Bryan, Willie, and I drove to the beach to see New Jersey's Coast Line. Mark opted to go to his room to relax and stated he would reconnect with us tomorrow. The Atlantic Coast Line was a beautiful sight. And after sightseeing around the city, we decided to grab a bite to eat then pack it in for the night. Riding back to the hotel, Bryan reminded me, "Remember you told me that you would go to the movies with me to see Transformers." "Yes I know" I responded. We set a date and time for the movies. Once back at the hotel around 8:00pm I was relieved when I was reunited with my luggage which had been delivered by the airline and left at the front desk.

Day three and another training day was drawing to a close when Bryan reminded me again about the *Transformers Movie*. I assured Bryan I was still on for the movie. The first part of the day was laced with new material. However, the second half of the day was a review of the course and hands on training. Never the less, the final test was going to be on the morning of the 5th day. Therefore, I knew if we were going to get to Bryan's movie, it would have to occur tonight because tomorrow night I would be studying for the course final. Class concluded at 4:00pm so immediately after class Bryan and I headed straight to the theater and purchased our tickets. It was opening night for the movie and we wanted to avoid the rush.

Since we had time to spare before the movie was to begin, we spent our time sharing some small talk, dining and shopping. We went to the dining area and sat down after purchasing some burgers. "These burgers are pretty good but *Five Guys Burgers* are better" Bryan remarked. "On our return through the Washington Dulles Airport we must stop by *Five Guys* and try their burgers again," continued Bryan. After eating we headed downstairs to the clothing store, *Burlington Coat Factory*. We found plenty of good clothing for great prices, but my intention was to only purchase a belt. Bryan on the other hand had a lot more shopping in mind; he purchased several pairs of jeans. Before our shopping excursion was over, I found this one shirt, a white tee-shirt with a checkered design on the front. Just outside of the checkered design was a symbol of a *fleur de leis*. This is the same symbol that represents my beloved *New Orleans Saints*. So I purchased the shirt and a belt and was done. Later we made it back to the theater. We both commented on how glad we were that we had previously purchased our tickets because we avoided the long line and the madness of a new movie release rush. Bryan had a great time, and said that he was thinking about coming back the following night to see it again. I quickly responded to Bryan, "The movie was good, but if you come back tomorrow you will come back solo."

The fourth day of course instruction came to an end about an hour early. The final exam was tomorrow so we all went in different directions to do what we needed to do to study. Willie went for a drive, Mark went to dinner, and Bryan went back to the mall to return the jeans he purchased the previous day at *Burlington Coat Factory.* My plan was to study a while then fit in a workout.

After being in my room for about an hour, I mustered up enough energy and convinced myself to put in a workout at the hotel gym. I had been working out back in Texas three times a week, that is, until two weeks before I left for the course. While working out I watched a special on the TV about Farrah Fawcett who had died earlier that day. Right in the middle of that special tribute was a news flash that Michael Jackson had been rushed to the hospital in an unconscious state. I figured there was not too much to this news story and that everything would be okay with the Pop Icon as I continued to listen to the news report. Thirty minutes into my workout, Willie called me on my cell and told me that he thinks the news is reporting that Michael Jackson is dead; I responded, "I have not heard that he is dead, the cameras are still rolling at the hospital. I don't think he is dead." I finished my workout and went to my room. The phone rang as soon as I closed the door to my room. It was Bryan. "Michael Jackson is dead" he exclaimed. "Are you sure?" I questioned. "They just announced it on the news" Bryan again voiced. In disbelief I softly uttered to Bryan, "Michael was only a year older than me; I've been listening to his performances for most of my life." After a few more somber exchanges, Bryan and I ended the call. At this point I was a little numb. Michael was a little different, but he was probably the most talented singer, song writer and dancer combination of all time. The world had lost an irreplaceable talent. I went up to Willies room to let him know that what he heard earlier was correct, but Willie was so devastated that all he could say was *"Damn."*

On the last day of the course within two hours the exam and review was complete; everyone scored higher than 80% on it. We all received our course completion certificates and were released from the class. Bryan and I had discussed taking the train to New York to see a few sites, but since it was already noon time, we figured it would be best to pack and get ready for the flight to our next destination in our overseas preparation process.

The next morning we all checked out of the hotel and headed to the airport in New Jersey's for our flights to Atlanta's International Airport. Willie had a navigation system in his rental car so we all followed him. Bryan and I rode together, and Mark followed with his rental. After turning in the vehicles and checking our luggage, we went back to the *Five Guys* burger restaurant as agreed. As before the burgers were great! After finishing our meals it was time to make our flight to Atlanta. Mark, Bryan and I were headed to Fort Benning, Georgia for overseas preparation and processing at the *Conus Readyness Center* (CRC). Equipment, training, physicals and final clearance paperwork were necessary before one is allowed to deploy. Willie on the other hand had already completed CRC training and screening and was headed straight to Iraq. We exchanged email addresses with Willie, and said our goodbyes as we went our separate ways. This exchange reminded me of some of my past military assignments because the group was starting to separate and usually you never saw those people again.

CHAPTER 4

WAR ZONE PREPARATION

When we touched down at the Atlanta International Airport unlike the nice cool weather we had in the north east part of the country, it was hot. The temperature was right around 100 degrees on that clear and sunny late June day. We had about a five hour layover in Atlanta. Therefore, we decided to get a rental car from Atlanta instead of waiting the five hours. Our bags were already checked and would have to make the flight without us. Besides I had a rental car reserved for me at the Columbus Airport, so if I got it now, I would not need to get one once I arrived there. The drive only took about one hour and ten minutes. When the three of us arrived, we hand about three hours to spare before our luggage arrived on the plane, so we decided to go and get some dinner at an Applebee's restaurant. The meal really hit the spot and before we knew it, it was time to go to the Airport to get our bags. "I know my bags will be lost" Bryan said as he had said before. I responded "Bryan will you just stop wishing yourself bad luck." We got to the airport about 20 minutes before the plane touched down. All three of us had a reserved rental at the airport that was reserved by our new employer. But since I rented the car that we came down from Atlanta in, and I had to retune it to the Atlanta Airport Rental Company, there was no need to exercise my right to get a car. Mark went ahead and got his rental, but when Bryan went to get his rental they we out of them; in spite of the one reserved for me that I did not get. "I knew it, I knew it, this sort of thing always happens to me" Bryan voiced. I responded "Don't worry about it Bryan. We all are in the same hotel and are all going to Fort Benning at the same time, so you will always have a car at your disposal." However, this time Bryan could not complain about the luggage because everyone's luggage made the flight. We picked up our bags and headed to the hotel in which we were staying.

The next morning we all piled into Marks rental and headed to Fort Benning which is located in Columbus Georgia. Fort Benning is an Army Post with multiple missions. We were there to receive a predeployment screening from the Army's CRC. The job I was about to undertake in Iraq was considered a Danger Zone Job. Preparations for such jobs were made at the CRC. All personnel deploying to Danger Zones were required to pass a medical screening prior to deployment; this included all military, DoD employees and contractors. The center was divided into many stations which took at least 5 days to complete.

It was unreasonable to send anyone to that region with serious preexisting medical conditions that could not be treated with medication. Besides, the primary reason for medical services in the war zone were battlefield emergencies and not routine treatment. Therefore, a medical fitness screening was very much in order. Candidates were expected to already have a complete physical along with dental x-rays prior to arriving at the CRC. If a potential candidate arrived without a complete physical, the usual five day process could easily double.

In addition to the medical screening, there were other predeployment requirements; First Aid Training, shots, issuing of a Common Access Card (CAC), and equipment issue to name a few. There was also a vision test and if a candidate wore glasses they were required to have in their possession two pair of glasses. Passing a hearing examination was essential in a war zone and if the candidate did not meet the minimum hearing level, they did not deploy. After all, potential dangers exist in a war zone and the ability to hear the dangers was essential. The required gear issued was a helmet, gas mask, flack vest and other body armor. The total weight of the gear issued was about eighty pounds. If I had known I would have my already heavy gear increased in weight by eighty pounds, I would have packed much lighter.

My wife and I thought it would be a good idea to see each other once more before I departed for Iraq. Since I was scheduled to leave on July 4 (the upcoming Friday) she flew into the Atlanta International Airport on the previous Wednesday. I was glad to make the seventy minute drive from Columbus in spite of her visit being just for a few days. After picking her up from the airport, her presence made the wooded scenery of thick pine trees illuminate on the drive back with the trip seemly much shorter. After the return trip we later contacted with Bryan and Mark and established a eatery for dinner that evening. Applebee's was the selected resterant and we all met at 6:00pm. I introduced everyone and dinner went off very well.

There was a gap in our processing and we did not need to come in until 4:00pm the next day. So, determined not to waist our new time windfall, we all decided to go and visit the U.S. Army Infantry Museum. Bryan and Mark followed as Gwen and I lead the way to the museum in our rental. After arriving at the museum Bryan continuously talked about how the museums sights made him remember many of the times that he had while on Army active duty. In fact, Bryan was adamant about how well he shot his weapon and his past marksman qualifications. Oddly enough, there was an M-16 Firing Simulator located in the museum. Mark and I had little interests in firing the weapons with the simulator, but Bryan wanted to show Mark and myself up so badly with his shooting he purchased tickets for both Mark and I without informing us. Mark is left-handed and I am right-handed but shoot with my left. This caused the two much older lefties to go up against an eager to shoot right handed expert. Sixty year old Mark hit the target twenty times. I followed with nineteen hits; but our marksman, Mister Bass hit the target only fourteen times. Now you know Mark and I teased Bryan the rest of the way; from that point Bryan was given a new name. Bryan was now known as junior.

The next day as soon as all CRC processing matters were addressed, my wife and I drove back to Atlanta and stayed the remainder of the evening with her brother and his family. It was good to see her brother again and that evening went well. But later that evening the reality of having to leave aggressively sat in because I had to be on the 3:00pm flight bound for Kuwait the next day.

The next morning was Independence Day, and with travel time and the two hours that airports require prior to your flight, I had to leave no later than 11:30am that day. The only time in our married life that we could compare to our upcoming separation was when I was on active duty in the Air Force. I had to spend a year unaccompanied on a remote tour in Korea. We got through that time and we would get through this one. Time seems to always go by faster when there is not much of it and this day was no exception. When it was finally time to leave we gave each other a big hug and I then left with little delay. No reason to prolong a painful event; I was on my way to the airport. I got there without a hitch; her brother had given perfect directions. Once again I met up with Mark and Bryan at the Airport and we prepared for our transatlantic flight to Kuwait. I boarded the plane on the day that now marks and symbolizes the birth of Americas Freedom, the day in which the United States Independence was established 233 years earlier; July 4, 1776. Oddly enough the name of the war campaign in which I was going to support in Iraq was christened *"Operation Iraqi Freedom."*

CHAPTER 5

WELCOME TO KUWAIT

The plane seemed to hang in the air forever when it was time to land after a soaring flight of 13 hours and 20 minutes. That was the last time that thing seemed slow because once we landed everything sped up at the Kuwait International Airport. Both Mark and Bryan had already been in both Kuwait and Baghdad before. Mark was returning to his third assignment in Afghanistan and Bryan had worked on other contracts in Baghdad. They both informed me on what I should expect and I was glad they were there since I had never been in the Middle East region before. The first thing that stood out about this voyage was that this place was every bit as hot as I had heard. The temperature was lingering around 135 degrees. The sky was beautifully bright blue without a cloud in it. There was a very high humidity and the sun appeared to be closer and illuminate more than I had ever experienced.

Passport stamping caused us all to be herded into channels that had no way to avoid. There was a health questionnaire added to the normal documentation that questioned if anyone had flu-like symptoms. Randomly picked passengers were questioned about their present health status. On one such attempt the *Third Country National* (TCN) working at the Airport was pleading with one of the new arrivals from our plane; "Please sir if you have any of these symptoms please let us know." Even if a newcomer had the early symptoms it was best not to mention them because an overreaction would occur that would either quarantine the newcomer or at a minimum delay their processing. But I understood this was just Kuwait's desperate attempt to control the Swine Flu Pandemic from entering their country.

Everyone passing through Kuwait had to pay 3 Kuwaiti Dinar (KD) for a Work Visa that extended for 90 days. None of us had Kuwaiti currency but we were graciously accommodated with an exchange bank that was located a few feet away. Women dressed in burqas with the facemask raised (a black nun resembling garment) were throughout the airport. There was little comfort in having persons covered from head to toe present. But there was no time to get sidetracked on such matters because the business at hand required us to move toward the passport screening area; and from there we went to gather our luggage at the baggage claim.

Our next move was to get on the bus that would take us to Ali Al Salem. Curtains were drawn on the bus. My assumption was this was to conceal what groups entered the country as a matter of security. Ali Al Salem is both an Air Base and Army Post which serves as a staging location for entry into Iraq and Afghanistan. My luggage consisted of a personal duffel bag, an issued duffel bag that contained my issued gear, and one old suitcase with wheels. The wheeled suitcase helped quite a bit by lightening the load I had to carry while in the airport. However, the wheels on the suitcase were pretty much ineffective once we made it to Ali Al Salem because most of the surface was covered with rocks. Therefore, my three bags had to be dragged across the souring heat.

The shelters at Ali Al Salem were large thirty man tents. However, we finally got settled into a tent and thank God there was air conditioning. In fact, at about midnight that night I had to exit the tent for a bathroom break. As I exited the tent the wind was blowing. The air that blew against my skin felt like a hair dryer set on hot. We stayed in the tents for a second night. However, on the third day about noon, Mark informed me he had been in contact with Baghdad and that he and I was booked for a flight. Mark was headed to Afghanistan and my destination was Baghdad. "What about Bryan's flight?" I questioned. "You must have someone looking out for you because you have a Blocked Flight" Mark continued. "What's a Blocked Flight" I questioned? "A seat on a flight has been reserved specifically for you." "But what about Bryan does he also have a reserved seat?" I again questioned. "No, only you have a Blocked Seat. The others will have to fly Space A" Mark again uttered. A Blocked Flight was a seat on a flight that was reserved. Space A means "Space Available" with no seat reserved. *Space A Seats* were granted on a first come first serve basis with a sign up roster. Space A passengers were the lowest priority. I had no idea of how the reserved seat had occurred with me, but then I thought Bert may have pulled a string or two or maybe there is something to Bryan and his bad luck. We all went to the McDonalds on Ali Al Salem and had lunch as we started to prepare for the upcoming Iraq Flight.

CHAPTER 6

GETTING TO BAGHDAD

On our departure day we gathered our bags and headed to the flight departure area. Since my flight was supposedly firm, I was able to have my bags placed on the pallet for packing. Mark added his bags to a pallet that was being packed for his destination in Afghanistan. Our bags were palletized (somewhat similar to checked baggage in a civilian airport) and packed away for the upcoming flight. However, that flight got canceled and another was generated in its place. This change caused my bags to get to Baghdad probably sometime later. I still had a firm seat on the newly generated plane, but then I found out that the new flight was canceled. The other guys had the same sort of luck because all Space A Flights headed to Baghdad were also full. That's sort of the way things went with flights in a warzone, different flights changing based on destination and priorities.

However, a new flight was generated and scheduled for departure in about 3hours. I managed to get a seat on the newly generated flight and again Bryan could not obtain one. Now I was starting to give Bryan and his bad luck theory some strong consideration. "Perhaps it would be best if we were on different flights" I thought. The existing issue for me was my bags were palletized and on their way to Baghdad without me. The three of us decided to get some dinner and then hang around the air terminal until the flight departure time. There was an announcement in the terminal that stated "If you have a reserved seat and BIAP is your destination we are starting to board the aircraft." When this was broadcasted I did not move. Bryan then uttered "Chancey they are gearing up for BIAP." I then questioned "Is BIAP the same as Baghdad"? "Yes" Bryan responded. No one had mentioned that BIAP stood for Baghdad International Airport. I told Bryan thanks for enlightening me and I would see him when he got to Baghdad and then we parted. At about 9:00pm that evening I was finally on my way to Baghdad. An adage my father would often say when I was a child and even sometimes today came to mind at that moment. His words were "Be careful what you ask for, because you just might get it." I was finally getting what I had asked for; a job in the war zone of Iraq.

The flight into Baghdad was a slightly nerve ending experience. The biggest hurtle to get over was just not knowing what to expect. I had served in the military for twenty years, but never in an area that was classified a war zone. The only assignment that came somewhat close was when I was assigned to South Korea. But that skirmish was classified as only a conflict. These military flights were a mixture of soldiers and contractors. Helmets and bulletproof vest were mandatory to be a passenger of this flight; without them meant no flight. A United States Air Force C130 was the type of aircraft used to transport us into Iraq. After flying for about forty minutes, the Aircrew performed a series of turns and sudden altitude descents as we approached BIAP to reduce the risk of possible hostile enemy air attack; these flight maneuvers were known as Combat Landings.

Upon landing we were all herded to an open area and positioned in a very broad military formation. I was hungry, thirsty and tremendously tired. The bus schedule, food, water and bathroom instructions were yelled out, but in a shouting quagmire fashion; too much sporadic information too fast. After the instructions were given we were released. It was about 1130pm. Both Mark and Bryan had been here before, but they were all still in Kuwait at Ali Al Salem Air Base trying to get a flight out. They knew their way around and I was hoping to utilize that but they were not here. At this point I had completed the CRC training, endured a transatlantic flight and finally touched down in Iraq. I had made it, but after checking on my bags they still had not made it to Baghdad.

It was now after midnight and I was far worse than tired, so I went the first bus that was headed to billeting, any billeting. I would get some rest and sort out exactly where I am supposed to be the next morning. I asked the bus driver what the destination of the bus was? The bus driver informed he was going to Camp Liberty's billeting. I was not sure whether this bulleting was close to my destination or not because I had heard Baghdad referred to as BIAP and Victory. But after I landed I discovered that the other side of BIAP was Salter Air Base and Salter Air Base was where we had landed. Later I would learn that all of the different entities (BIAP, Salter Air Base, Camp Victory and Camp Liberty) were all separate locations within one big military installation located in Baghdad.

Then I saw an African American man about my age walking toward the buses. We exchanged salutations as he walked by and I asked him where he was headed. He answered "billeting;" with my exhaustion level as it was, that was the only right answer. At this point I did not care whose billeting I rested in so I climbed aboard. I was so tired that anywhere that had a soft bed and a pillow was getting claimed by me. He and one other guy, this time Caucasian were the only other passengers on the bus in those early morning hours. The driver of the bus was a TCN.

As we left the Salther Air Base's area, there was a drastic change to the terrain. The Airport and its immediate surroundings were neat, trim, and for the most part aesthetically pleasing. But once we exited the

Air Force Base area, the streets had pot holes everywhere. There were bullet holes in the approximately fifteen foot cement walls equipped with barbed wire on top of them. Without any doubt a war had occurred here. After additional dialogue had been established with the other two passengers on the bus it was conveyed that both of the gentlemen were about three years older than I and both of them had been working as contractors for about four years. The African American guy was from Fort Hood Texas and the Caucasian guy was from the Dallas area. I of course was also from the same state. We all thought it was a little humorous for only Texans to meet up at that hour in a desert terrain much more harsh than any desert terrain of Texas. We went about five more minutes down the road and the Dallas Texan got out. He quickly disappeared behind a concrete barrier with a design that was unknown to me; later I would find this barrier was called a T-Wall. Basically, a T-Wall was a concrete upside down T that was about two feet thick and about twelve feet tall. The T was inversed so that the wall would stand. The T-Walls function was to serve as a protective perimeter that prevented a possible enemy from having the capability to do drive-by's. We rode for another ten or fifteen minutes and finally arrived at Camp Liberty's billeting.

Billeting was pretty much the same in Baghdad as Kuwait; a *"thirty Man Tent."* Cots, not beds were the sleeping apparatus and the tents were again air conditioned. My new associate showed me where the recreation center was so that I could make a call to my new employer the next day. He also gave me a heads up on checking your bottled water very close because sometimes they appear to be tampered with. Since my luggage and I made different flights, the only clothes that I had I was wearing. So needless to say, that night I went to sleep a little sweaty. We talked for about two more minutes on some of the do's and don'ts, of the installation and the area and just like that I was out.

The next morning I quickly noticed one big similarity between Baghdad and Kuwait; they both were Hot. The temperature was souring around 125 degrees and again a blue sky with no clouds. The guy from Fort Hood informed me of the location of the Post Exchange. This was well received information because the one pair of clothes situation demanded urgent attention. Soon afterward we wished each other well and went our separate ways. I then went to the recreational center to make a Defense Information System Agency (DSN) call to my friend Bert. I called to let him know that I had finally arrived in war zone. The conversation went along these lines. "Smoooth, I have finally made it over here" I voiced. "Bert replied" Smoooth is that you?" "Ya man it's me" I responded. "Does it look like I will be getting to come that way" I questioned" "I have been working on that as best as I could, dropping hints and whatnots" Bert responded. Bert had already contacted the headquarters people and mentioned that he wanted to have me assigned to his sight. Bert also informed me that I would need to talk to Kimberley who was the head admin clerk because she is a clerk that sometimes had some input on personnel. Bert instructed me to ask Kimberley the following: "What are the possibilities of me getting assigned to Warrior Site."? I thanked Bert for getting me on board with this firm we hung up and I immediately called Kimberley.

I contacted Kimberley but immediately received a very negative reception based on her tone. In fact I did not know how to place her. She was not quite rude, but at the same time she was clearly not warm or friendly. She was abrasive and to the point when there was no need to be. Considering she was the voice that represented the company, I wondered how she kept that job. However, I did conclude that she was someone that you did not want to get on the wrong side of. I asked about the possibilities of being assigned to the Warrior Site, but no answer was provided and the negative tone continued. At that instance it dawned on me that of the brief time I had been in Iraq, I had met who appeared to be an enemy. I had done nothing to foster that relationship, but this enemy was on my team and seemingly posing as a friend.

The day had progressed to about 10:30 am and there were several other things to do before sundown. My next move was get in touch with my upcoming work center. I called my new work center to inform them of my in country arrival. This one call turned into a circus act of many changing hands. There was voice after voice projecting out over the phone asking questions about who would come and pick me up with no one offering any answers. Finally a pleasant voice greeted me on the phone. "Hello, I am Alfred, the Site-Lead; I will be coming out to pick you up. Exactly where are you located?" he questioned. Alfred's voice sounded similar to the actor John Wayne, but at a slightly higher pitch. "I am at Camp Victory's billeting" I responded. Alfred then questioned "Are you on the side with the two poles and the big antenna, or the side near the entrance?" "At present I am at the recreational center located on the back side of billeting. I don't recognize the landmark you've just described," I responded. "I know where you are" Alfred replied. Within ten minutes Alfred arrived at the central point near billeting where we agreed to meet. Alfred arrived in a small white Mitsubishi extended cab pickup truck. He was a Caucasian male who appeared to be about my age and height and weighted about 200 pounds. "Welcome to Baghdad" he voiced. "Thank you" I replied. "How was your flight?" Alfred questioned. "The flight was fine, but my bags did not fare as well, I'm hoping to reunite with them soon" I replied. "Let's get your lodging squared away so that we can start getting you settled. I think we should be able to get you into the Bunkhouse" Alfred continued. Alfred and I arrived at the Billeting office. At check in, I was issued a pillow, blanket and bed sheets. Once I was all squared away with my lodging, I secured my combat gear in a locker at the bunkhouse. Alfred then took me to the work center.

The work center was highly active and booming with maintenance activities. While being oriented around the center, I glanced up and noticed right in the middle of things was Venture Willie working on top of a truck. I will admit I felt a sense of comfort to see a familiar face. We extended greetings to each other. Willie informed me he had already been there for about a week. We made plans to catch up with each other later as Alfred and I continued viewing the work center.

I couldn't have been welcomed any more cordially than the greeting that Alfred provided. He continued the orientation and informed the facility is a regional hub that sponsored most of the training for the WarWizard program. It encompasses four offices; the office of the Facility Commander, a Navy 05 and his assistant a Navy E7, the second in command Department of Defense (DoD) Lead, the third in command the Facility Site Lead (Alfred) and the Lead Supply Manager also maintained an office in this area (Walker). As we toured on, I observed the facility had three bay doors with enough room to house up to six large military vehicles at the same time. There was a back office for Field Service Representatives (FSRs) which were the workers who provided the bulk of the manpower and three other offices on the other side of the building that the equipment specialist worked from.

We made our way to the break room. In order to get to the buildings two bathrooms a person had to go through the break room. My stomach had been talking to me every since Alfred picked me up so I was elated to see the table in the break room that housed foods such as pop tarts, cereals, oatmeal packets, juice packets and other snack foods. It was there for the taking. As I moved in closer I quickly came to the conclusion that even though I was hungry, there was no way that I was going to touch it because it looked as if it had been there for months. There were layers of what looked to be dust and dirt that had formed on it. I had a good laugh to myself when I found out later these items were actually fresh but often fell prey to daily dust accumulation. Also, it was explained to me that it was necessary to sweep the floor and wipe dust off items two and three times each day and even more on windy days. Let's just say keeping the snacks looking presentable was a job in itself.

After making our way through the break room, where most of the Field Service Representative's (FSR) generally reside during times between jobs. Alfred appointed a guide to me to aid in the things that I needed done. Brenda was an African American girl who was about thirty-five years old. She asked me how she could best help me in getting settled. I explained that it was really past time for me to get something to eat. Brenda inquired "Would you like to watch a movie while you eat?" I responded "Yes." Therefore, we ventured to the dining facility (DEFAC). As we approached the entry point of the DEFAC, I observed individuals being searched after they presented their identification cards to Armed Ugandan Soldiers. Brenda informed me that prior to entering the DEFAC we would have to show our identification card (Common Access Card [CAC]) to the Ugandan Soldiers who were securing the facility's entry point. In addition to presenting an ID card, all TCNs were required to be searched prior to entering the DEFAC. Ugandan Soldiers also secured and controlled the entry points of the major facilities throughout the compound to include the gym, base exchange and other common large personnel congregation entities. A line of about eight people were being searched prior to entering the facility. I filed in at the back of the line but Brenda motioned to me and we simply went around the line, showed our card to another security soldier we and entered the facility. At the moment I thought this action was a little odd and perhaps even a little prejudice but that was what occurred because we had a different clearance as Americans than the TCNs. Also, in additions to Americans this practice was extended to British and Australian contractors. Later it was made known that the Ugandan Soldiers also controlled the security at the gates and the perimeter of the base.

Most of the DEFACs throughout the large complex were managed by the Army but the DEFAC closest to Salter Air Base was managed by the Air Force. The Air Force's dining facility was smaller with many comforts not existing in most of the other DEFACs according to Brenda. It was basically a serving line / buffet-style restaurant combination. Surprisingly the food choices were plentiful. There were four or five different meat choices, an assortment of breads, vegetables, fruits, and about twelve different dessert choices. And to top everything off, there were also approximately ten different ice cream flavors. No money changed hands; the method of payment was your CAC Card. With each visit you could go through the line as often as you desired as long as you did not exit the facility. However, if you did exit the facility, the rules granted you two more entries for a total of three separate entries each day. One could use their CAC Card for breakfast, lunch and dinner or three lunches or any other combination as long as there were no more than three scans each day.

After getting lunch I had to take care of getting some new clothes because of my lost luggage issue. Brenda took me to the Post Exchange and I purchased two pairs of work pants, a ten pack of gray tee-shirts, underwear and socks. My thoughts were to purchase as minimally and cheaply as possible; just enough to hold me until my luggage and I were reunited. After purchasing the clothing items, the change I received back was a mixture of regular American coin currency and Pogs. Pogs were a circular coin shaped AAFES paper Gift Certificate. The front of the Pog Coins had a value of 5, 10, 15, 20 or 25 cents. The back of the Pog usually had a picture of a uniformed military member or military equipment and sometimes an American Sports Champion.

After my shopping detour was over, Brenda strongly encouraged me to put in my leave request as soon as possible. In addition, she instructed me to the location of the DSN phone connections near the living quarters so that I could call my family and let them know I'd arrived safely. This was the most valuable information of the day for me because from that point I called my wife every night at least once until I acquired my own communication methods.

Brenda and I returned to the maintenance work center. From there I resumed my orientation with Alfred. He suggested I take the remainder of the day off in order to rest up from the flight. At the same time, he informed me of the 7:00 am work show time for tomorrow. This was obviously a strong management move based on his experience, because after getting a good meal in after what I had been through, I was good for no one. All I wanted to do at this point was to make my way back to the Bunkhouse for some much needed wind down time. I thanked Alfred and Brenda again for all of their insight and direction then headed back to the Bunkhouse. Later that night at the Bunkhouse, as agreed upon earlier, Willie and I spent a brief time catching up. Willie's bunk was on the other side of the room directly at the foot of my bunk. Due to my fatigue and jetlag, we agreed to get together again another day and shortly afterwards, I was asleep.

CHAPTER 7

THE ELECTRONIC BUBBLE CENTER

The distance to work was only about 120 yards therefore getting to work was only about a five minute walk. I arrived at about 10 minutes until the hour. The members of the facility were already gathering in the bay area in anticipation of the morning meeting. At 7:00am sharp the meeting started. Alfred introduced me and two others as the newest arrivals. The meeting progressed on for about five or six minutes with routine business agendas. At the conclusion of the meeting, everyone disbursed into different directions leaving only about twenty percent of the shop's personnel. In the moment I observed with wonderment. I had no clue what was occurring until Brenda clued me in. I learned that some of the workers went to take or pickup clothes at the laundry while others went to breakfast or both. All meals were free along with laundry services or at least it was free to the contractors because their sponsoring company picked up the bill.

As mentioned earlier, Alfred was the Site Lead, but the persons managing the floor were Brenda and Daniel. Daniel was a Caucasian male that stood about five feet four inches tall. He was fifty-five years old with gray thinning hair. Daniel was a talented technician and a very religious man, based on frequently reading his Bible during slow maintenance times and seemed to also live a moral life. His title was Floor Lead. Brenda was the Quality Control (QC) Inspector. Our mission in the WarWizard program was to install and maintain electronic jamming devices in military convoy vehicles. We also performed maintenance check and modified equipment when ever needed. When installed, our equipment placed an invisible electronic protective bubble around the vehicle in which it was installed. This was due to our methods impeding upon the effectiveness of enemy road side bombs ability to explode. Often when military convoy vehicles were on patrol and move through neighborhoods, our jamming equipment often knocked out phones TVs and radios. The range of the shielding bubble was classified, but the distance was great enough aid in a much greater number of troops returning home safely. We were a 24 hour coverage facility with the two 12 hour shifts starting and ending at 7:00am and pm. On the day shift, Brenda or Daniel inspected all work performed prior to any vehicle leaving the facility.

My position title was Field Service Representative (FSR). There were about thirty-five Field Service Reps at the facility. My best guess was that there were about sixteen FSRs that were permanently assigned there and the rest were in-bounds, out-bounds or transients. Alfred ran a tight, but professional ship and the success level of the WarWizard shop in Baghdad showed it.

Later that day at the work center I met a guy who seemed to be a decent fellow. He had been on the project for four years. His name was Richard West and he was also an FSR. We both were in our late forties, (he was forty-eight and I was forty-nine), married for over twenty-five years, our wife and children were the same age, and both served in the military and were from the south. But we had our differences I was a retired Master Sergeant from the Air Force after serving for twenty years. Richard, on the other hand had served in the Army for two and a half years and for his family it was his wife who was a retired Army Sergeant First Class. Later he would explain why he had such a short and odd length of time of military service. Richard and his wife had been in the WarWizard Program for four years. Both Richard and his wife Bernice started as FSRs but she was removed from the maintenance facility and moved to the Headquarters Office with Kimberly. In the days to follow we continued to exchange life events in the break room when there was downtime. Richard mentioned to me several times that he had a BMW back in the states that he could not wait to get back to driving when his vacation starts.

Four days later Richard arrived at the Bunkhouse at about 500am. Then Richard uttered "Willie you need to get up." Dazed in the middle of sleep, Willie responded, "Get up for what?" Richard continued "You need to pack and get your things together." Willie then quickly got a clear head and voiced "Richard what are you talking about?" Your flight for FOB *(Forward Operating Base)* Warhorse leaves in an hour" Richard continued. "No one told me about leaving and having bags packed today, who was responsible for letting me know?" Willie responded with mild anger in his voice. "I don't know" voiced Richard. "Well who told you to come and take me to the airport?" Willie again questioned. There was a twenty second pause and the Richard said "Bernice." Willie continued to grumble about the way the events unfolded. But he got packed and made the flight in time. I cannot confirm which one of the West twosome dropped the ball with Willie but based on the extreme difference in each of their charter I have a strong opinion on the matter. But as different endeavors continue to occur throughout this journey, you will be able to draw a reasonable conclusion as to who did not provide ample notice for Willie. Three more days passed and my new found friend and his wife Bernice were on a plane back to the states for his three and a half week vacation.

Training started almost immediately as many trucks arrived to the shop for maintenance. Baghdad was the location of Headquarters for the WarWizard program. In addition, Baghdad served as a hub training facility. When new personnel arrived to the program it was Baghdad's responsibility to give them that ground floor training and then send them on to their *Contingency Operating Base (COB).* If an FSR was selected to remain as part of the *"Baghdad Facility Maintenance Team,"* permanent tools were issued to that person and they remained in Baghdad at the Baghdad Facility. FSRs selected for other COBs would return the tools issued for maintenance the day before their departure. In my case, permanent tools could not be issued to me until the question of my permanent status was answered.

Assigned as my initial trainer was a sharp 32 year old Puerto Rican FSR from Louisiana named Carlito. He showed me ground floor maintenance on the Jamming Systems. In spite of our age differences, we hit it off well. Perhaps our harmony was so quick in part because we both liked the same football team; the New Orleans Saints. In fact there were more Saints fans in that facility at that time than any other team. To my credit I jumped right in being attentive to all that training and quickly learned how to install the systems and troubleshoot maintenance problems. However, one issue that interfered with cohesive training was other FSRs asking to borrow tools. This may sound trivial, but often the person that borrowed the tool would not return it and with so much borrowing going on sometime the person loaning would not remember who the borrower was. All in all we continued through the training adapting to the tool situation. Dealing with this unique tool situation in training and what seemed to be a lack of experienced personnel, reminded me of Richard's absence. I made my sentiments known by commenting, "It will be nice when Richard returns." However, whenever I would mention this to some of the veteran FSRs, I would get comments like "Well I don't miss his loudness" or "Haven't you noticed that it is quiet since he left?" I did not understand why they all seemed to be so critical of my new found friend.

About the third day of training, Alfred introduced Bryan Bass to the group at the morning meeting. "So you finally got out of Ali" I laughingly stated to Bryan. "Yeah man and it was a nightmare" Bryan responded. "We had two flights cancel on us. And the third one was just about full, but I was next to the last guy to make that flight. I have the worse luck" Bryan complained. "You really need to stop saying that Bryan, besides your luck actually was good from the prospective of *you did make the flight*" I voiced. At that moment Alfred interrupted our small talk to inform us that we had an *in processing* appointment with Kimberley tomorrow at 10:00 a.m. He stressed that it was important that we be on time. I thought to myself that I would finally get a chance to meet the rude administrative person that seemed to will so much power.

Lunch at the dining facility presented some odd rules that day. I walked into the dining facility with my sun shades positioned on the top of my head as usual. After signing in by scanning my CAC Card, I was halted by one of the soldiers who worked there. The soldier told me that I could not wear shades on the top of my head in the facility. Puzzled by the instructions, I questioned, "Are you serious?" "Yes I am" the soldier responded. Therefore, I moved the shades from my head to the front ridge of my shirt collar and started to proceed on. Then the soldier stopped me again stating "there can be no items worn above the waist other than your identification badge in the dining facility." This seemed completely senseless to me especially since this had not been the first time I had arrived with my shades positioned above my waist. I then stated, "This seems a bit strange. Why do I need to put my shades at waist level?" "Soldiers cannot have items hanging off their uniform above the waist, and we are holding contractors to the same rules as the soldiers" replied the soldier. My initial thought was *"now this is really stupid."* But staying focused on the reason I was there, I did not let a matter such as that get me off tract. Therefore, I without further question complied and put my shades in my pocket and moved on. As I walked away, that soldier stared at me as if I had killed his best friend while they were on patrol. Reflecting on the matter later, I understood that the key point the soldier was making was they could not adorn certain attachments while in uniform. Being a retired military member myself, I clearly understood the rules that the service member must abide by while in uniform. But, contractors are not in uniform. Our proper attire was regular civilian clothes with our badges above the waist.

Camp Victory was the major component of the Victory Base Complex (VBC) which occupied the area surrounding the Baghdad International Airport (BIAP) and several other camps that existed within its boundary (the wire); This included both Camp Victory and Camp Liberty. This massive single entity made up the Victory Base Complex. With its combination of Camps within a single boundary think of the Victory Base Complex as a metro area. Each of the camps possessed at least one DEFAC, but Camp Victory was the only one with such a blindly enforced rule. However, it would appear that since my arrival, for the second time I had made and enemy without attempting to do so. Surely this soldier had an enemy outside of the wire in which he can engage; after all, we were in a war zone.

CHAPTER 8

A HEADQUATERS ROCK THROW

Alfred assigned Fernando, (the training manager) the task of getting Bryan and me to our in-processing appointment with Kimberley. Bryan and I were in place and awaiting our transportation by 9:35am for the 10:00 a.m. appointment. The drive was only about 7 minutes long which allowed us to arrive about 20 minutes early. Fernando dropped us off in front of a building, instructed us to call him back when we were ready for pickup and drove away. We discovered that Kimberley's office was behind a coded fence that restrained us from entering the facility. Neither one of us had cell phones to call any of the persons inside for the code. Therefore, we decided we had no choice but to pick up rocks and throw them on the top of her tin portable building in hopes of getting her attention. After about three tosses, Kimberley came out and let us in. She was an African American woman who appeared to be about forty years old. I guessed her stature to be about 5ft 4inches tall and about 165 pounds. The only words she managed to muster up to us for the next 20 minutes were a stern "You are early, have a seat and I will see you at your appointment time. She was not polite, poised or professional, just direct. So for the next 15 minutes we sat and waited. During that time a third newcomer showed up at about 5 minutes to the appointment. The newcomer's name was Maurice. He was a retired Master Sergeant from the United States Army. Maurice was an African American male that stood about 6 feet 2 inches tall and weighed about 220. I viewed his facial expression as somewhat odd and slightly goofy. Conversing with Maurice caused me to re-evaluate my previous judgment of someone who appeared to be goofy to quite intelligent. He was in the supply career field. I had met him briefly back at the screenings that were at Fort Bennings and to my surprise, Maurice lived two subdivisions away from me in Texas.

Promptly at 10:00 am Kimberley stopped what she was doing and started our in-processing. After about 15 minutes of filling out forms Kimberley informed us of the different locations we would be permanently assigned. Bryan wanted to stay in Baghdad and I wanted to go to FOB Worrier. Then Kimberley announced the locations we each were about to be assigned; "Maurice you are going to FOB Tallil, Bryan you are going to FOB Worrier and Kingston you will be staying at Baghdad" "Kimberley, is there any way that Bryan and I can swap, because he would like to stay here and I wanted Worrier?" "Oh, you wanted to be with your friend Bert; well we don't send friends to work for friends because friends don't work, so no you two cannot swap" she replied with a sassy affect. Now I never saw that coming. She did not know me from Adam yet she made this huge assumption about me. I couldn't help thinking that this position of power was quite unusual for an administration clerk. Clearly this was not a position suited for anyone who would make such an assumption then make a decision based on that assumption. According to my conclusions, there were no more questions concerning the character and demeanor of this individual. It was now clear this clerk had an attitude problem, but in addition to her attitude she seemed to have already classified me for whatever reason as someone or something less than a professional. However, that was okay because of her rudeness and snappy judgment, I had done the same with her.

Witnessing all of this interaction at the other end of the room was a cute tiny dark skinned African American girl named Ginger. She was about 4 feet 11 inches tall and no more than 100 pounds. I was under the impression that Kimberley was the lone star of that office so I wondered what Ginger's role there? At exactly two hours later Kimberley voiced "this concludes today's in-processing and we contacted Fernando for pickup.

Later I learned that Ginger was actually assigned to the Kuwait WarWizard Shop; however she was the designated person to fill in for Kimberley because Kimberley was about to leave for her 3 week vacation.

This in-processing went on for 3 days. Each day we spent about 2 hours with Kimberley. Day two of processing started with Maurice telling Kimberley she needed to speak to him in a more professional fashion. Kimberley responded to Maurice in a rude and argumentative manner. She then motioned a negative hand jester to Maurice. She spared us all the tasteless use of *the finger,* but rather an *"I discount what you are saying hand wave* followed by a verbal *"Whatever."* To make matters worse, her supervisor Herbert Scott, (the programs Deputy Program Manager) was sitting no more than 8 feet away from the episode. At no time did he intervene or even acknowledge our presence or the incident. He only quietly sat there at his desk staring at his computer screen. During this time the number one person of authority in Iraq, the Product Director for the WarWizard Program, Rusty McConnell was TDY in Afghanistan. I wondered if he had witnessed the actions of Kimberley would he have responded differently.

CHAPTER 9

BAGHDADS REVOLVING DOOR

Dealing with someone as unprofessional and witchy as Kimberley made for a refreshing break to have met someone on the total opposite spectrum as Brenda. She was my initial guide when I arrived and now she was within two or three days of permanently returning to the states. Brenda had worked two years in Baghdad and had previously mentioned that she wanted to get back to the states to finish up education and start investing in real-estate. I had a very professional impression of her and really hated for her to exit. Two days before she left, she divided up her tools among her friends, but I couldn't help feeling a slight preference when she gave me her special tool. Brenda designed the tool and had it crafted in the welding shop. It was a screw driver with an Allen Wrench welded to the end of it. It was quite a handy tool when installing systems because the Allen Wrench was commonly needed on the systems. As tradition would have it, on her last day, there was a gathering at the end of the day but this time it was time to send one of the good FSR's away properly. The Facility Commander was a Navy O5. He said some kind words honoring Brenda's contribution to the mission and presented her with a hand sized model of a T-Wall that all of her friends and associates had signed. The Site Lead Alfred also joined and expressed some strong positive words on her behalf. After this period of accolades, Brenda finally relayed a brief expression of gratitude. I wished her well and hoped she wouldn't fall prey to that old cycle of *let's keep in touch but never do* syndrome; but as anticipated that was the last time I heard from Brenda.

Within my first fourteen days there were several inbound personnel who arrived and several to leave. My good friend Bryan was one of the ones to exit COB Victory. Bryan was granted the opportunity to head to the same FOB Worrier site in which my good friend Bert was Lead Supervisor. The other personnel were either being sent to Afghanistan, back to the states or being sent out to different FOB's after receiving initial training at the Baghdad Facility. In the mist of all the personnel changes was the constant personnel request for FSRs to assist the WarWizard Facility in Kuwait. On two occasions earlier I had heard of personnel shortages in Kuwait that required four or five additional maintenance personnel. Today the rumor was the WarWizard Facility in Kuwait needed ten of the Victory Facility's workers for temporary duty for 60 days. Later during the end of the day meeting. Alfred confirmed that the ten person rumor was indeed true.

Camp Victory Facility was a regional hub that sponsors most of the training for the WarWizard Program. In addition, Headquarters for the WarWizard Program was also located on Camp Victory which is the same office where Kimberley worked. And everybody's coming and going had to come through Kimberley's office which was quite a face for this franchise.

I remembered Brenda encouraging me to put my leave in quickly because the leave program had many imperfections. I started to see what she meant because the flaws with the leave program became apparent when a change had to be made with my leave. The leave policy required the entire leave form to be resubmitted whenever a change was required. Therefore, back to the headquarters office I went to resubmit it. Herbert Scott was manning the office by himself when I arrived. Through our dialogue he informed me that he notified everyone of the new changes to the form were on their way. As I turned in the changed form, I brought to his attention that there seemed to be an error in the leave form because no signature slot remained for the person requesting leave to sign. Herbert's response was "When I created that form my intent was to leave it up to the supervisor to sign it, not the person requesting the leave." I hesitated for a moment because I was completely floored by his seemingly illogical response. I thought I must have heard him wrong because I had never known this step (the requestor's signature) to be excluded from any organization's form. I have been a part of the military, the DoD, and the civilian realm and never had I heard of such an unprofessional position. Realizing that I didn't need to make any enemies in this office, I backpedaled from my position by stating in a joking manner "That's a little unexpected, but we are in Baghdad." Since Herbert had no clue as to why a leave form should have a location for the requestor's signature, I saw no reason to engage any further on the matter. There was no reason to make an enemy out of the guy who was the second highest ranking manager in the program within the country. However, the explanation Herbert provided strongly suggested he did not possess the proper skill set for his position in which he held.

Work days were long and arduous at times. A regular work week consisted of 12 hour days, 7 days a week. Trucks constantly came in for maintenance all day every day. Constantly trucks were scheduled for maintenance checks to ensure that they were working properly before they left the COB for a patrol. Others arrived that needed a complete systems install. It would take a team of three personnel about 2 hours to perform a complete install from scratch. Sometimes trucks came in to have all of our sensitive equipment removed in order for them vehicle to be either shipped back to the states, or transferred to the Iraqi military or police. Some days we pushed out over forty trucks through the WarWizard Facility. As a stress reliever, when the work was done or in between impending truck deliveries we would often toss a football around inside the bay area.

Our work area resembled an airplane hangar with 3 large bay doors that remained open at all times. The heat of the region and the constant flow in and out of trucks made it difficult to regulate the heat inside the facility. If it was 135 or 140 degrees outside, it was close to the same inside. Therefore, it was important to drink at least two and a half gallons of water each day to avoid the risk of getting dehydrated because in those temperatures water exited our bodies fast.

Work clearly had its importance because that was what we were there for. Our work schedules gave us little time for leisure. Just my short time there created an awareness that screamed I had to do whatever it took to keep my sanity intact. Being that I considered myself a conservative, no nonsense family guy, I figured it wouldn't take much to keep things on an even scale. My wife and family are my world and my refuge, so later that evening I called my youngest brother on an old cell phone that Bryan had given me. We conversed about the events that I was experiencing thus far with this new position. After speaking with him I spoke with his wife, my sister-in-law. She mentioned to me a few of the news reports she had heard and expressed her safety concerns since I was in a war zone. I set her mind at ease by letting her know that it was a dangerous area but the risk is diminished to a degree as long as you follow procedures. In addition, she told me that their youngest child (my youngest niece, age 8) that I called Soggy Bottom was very worried about my safety. Since it was past Soggy Bottoms bedtime I told Eleanor to reassure Soggy Bottom that I was in no major danger. We exchanged love and well wishes and concluded our call. As part of my usual nightly routine, I called my wife. I can always count on her to keep me going by reminding me of my worth to her and making me feel this sacrifice really mattered in more ways than one. I concluded my round of reaching out by calling my parents. I spoke to my mom briefly then to my dad. It was not easy for my mother to digest the fact that one of her children was in a war zone. But my father, a former military man himself was able to approach the situation much better. Usually conversations with my dad are never brief, and this time was no different. Forty minutes had lapsed and we were still full stride into our conversation. The local time in Baghdad was about 12:15am when I heard automatic machine gun fire in the distance. I told my dad that this was the first time I had heard that. But this was also the first time I had been outside at that hour since arriving in the country. I concluded that the gunfire was probably off base and posed no threat to the base. Since it was getting late and I had to get up early we ended the call. I was able to rest peacefully that night having talked to my family.

Ten days into my home away from home stay I exited the bunkhouse and was finally issued a permanent room. The living area was a 12 feet x 11 feet portable that I shared with another contractor named David. Our small metal home was air conditioned of course with a small dorm sized refrigerator, two twin beds and two wall lockers. The wall lockers were used to make a partition to divide the room. My roommate David was about ten year's younger, slender build, Caucasian with an incredibly laid back demeanor. David had been contracting for three years. He was the perfect roommate. We found it easy for us to stay out of each other's way; perhaps because David's routine was to go to sleep every night at about 8:00pm. As for me, I usually stayed up until midnight watching TV or a movie on my computer. I handled my electronics in a silent fashion and there was never an issue.

A new day at the job presented with a new course of action. Alfred echoed the new rules from Headquarters at the morning meeting. Alfred began rattling off the new guidelines stating, "From this day forward contractors will no longer leave after the morning meeting for breakfast. If you want breakfast you will need to get it before you get to work. If you need to pick up laundry you will need to pick it up during your lunch break. This applies only to civilian contractors and not to DoD Civilians." These stipulations made it clear to everyone who was DoD and who was a civilian contractor. On the same token it was also made clear that all shop personnel were not CSS employees. In fact there were representatives from at least eight other firms in the facility working as subcontractors to the primary contract who was CSS. I'm not sure why someone would arrive at such a divisive decision; it did nothing to support the *one* team, one fight mindset. It appeared that most employees worked well together and that no one cared who got breakfast or not or when. Besides, ninety nine percent of the work was always scheduled after 8:00am. Even if a vehicle came in unannounced that needed our attention there were always enough personnel on hand to handle the mission. But those were the new rules and I wasn't one to defy them.

Before that meeting, I had already befriended several of the DoD Civilian personnel who usually went to breakfast or the laundry after the morning meeting so when their van took off after the meeting, I usually went with them. So for now, that camaraderie could only occur at lunch time. For the first time (or at least since I had been there) the Deputy Program Manager Herbert Scott was hanging around the facility during breakfast time. He continued to be in the area at breakfast time from that day forward to ensure the new rules were not violated. Later that same day my spirits were lifted when Alfred told me that my leave had been approved. I kindly asked Alfred for a copy of the approval form in which he had just conveyed with his mouth, in which he assured me he would.

CHAPTER 10

A LOT TO LEARN

The next day we again had our share of new arrivals. A fifty five year old African American male named Jeffery Jameson was one of the newcomers. Jameson was very dark skinned and stood about five feet and five inches tall. He appeared to weigh about two hundred and twenty pounds with most of his weight centered on his waist. His overall appearance was that of a somewhat unhealthy individual. Another highly distinguishing characteristic Jameson possessed was ear splitting loudness. When I say loud am not saying vociferous, I mean his speech is just plain loud. Whenever we'd sit in the break room we could clearly hear him speak through the door of the break room when he was at least sixty feet away.

Jameson and I did not hit it off right away. We quickly clashed one day when Jameson heard me giving my diagnosis of why the company vehicle air conditioner wasn't working. Without any basis, Jameson boldly tried to disclaim what I said. I responded to him by stating "I have gone to air conditioning school and worked in that field for ten years. Since age nine, I trained in the field with my dad who owns an air conditioning repair company. So now I have a question for you; *"What component of an air conditioning system does the compressor pump directly into?"* "The evaporator" Jameson responded. I then replied "That is not accurate; air conditioning systems pump directly into the condenser." Next time make sure you are educated on the subject before you disclaim or challenge what a person is saying." Jameson acknowledged my competence and we were able to move on from there. Ironically, Jameson and I found a good happy medium because we later discovered we both were diehard New Orleans Saints Fans. So I guess it was safe to concede that the love of the Saints caused that situation to end on a positive note.

About fourteen days into my stay Malcolm Brown arrived; the actual person assigned as the Floor Lead. I say *actual person* because when I first saw him I thought he was just another newcomer. This was because I was under the impression that the floor was ran by Daniel and Brenda because that was all I had seen since my short arrival and until now, no one had indicated anything to the contrary. Malcolm was returning from emergency leave because his father had passed. Malcolm was second in command right behind Alfred. He was an African American top notched forty year old leader who knew the secrets of motivating personnel. He was about the same size as I. I am sure he probably learned most of those skills while spending twenty years in the United States Army; he was a retired Sergeant First Class. Basically, Malcolm put to practice the management example of *being friendly, firm and fair* when dealing with personnel. Once a leader has mastered the art of when to apply each technique, the motivation portion of the supervisor has been conquered. Malcolm performed this practice so well that it was as if he wrote the book on motivation.

The following day Malcolm gathered me and three others and informed that we would need to go to the Palace to complete the final part of the process to obtain courier badges. I then questioned "Malcolm, what is the Palace and where is it"? "Malcolm then responded "the Palace is one of Saddam's old Palaces. It is now used by the General as a Headquarters. All you guys need to do is go there take a picture for your ID card and you should be back in about forty five minutes."

The next day we ventured to the Palace and this place was quite a spectacle. The Palace was located on a lake with seemly no expense spared. After getting through the security, the first vision that captured my eyes was a humongous chandelier hanging from the ceiling. My impression of this place seemed as if I was in a museum. Tours did occur, but the twelve hour shifts that we worked each day would never permit any of us to partake in them. We were tempted to take a sample tour, but the traffic load to get our pictures and then the Courier Card did not allow it. One hour and 30 minutes later we returned to the WarWizard Facility.

Toward the end of the day there was a large trail of black smoke departing into the sky. I had no idea what was the cause; consequently, I popped out my camera and took a picture. Daniel had been in Iraq for several years and witnessed me taking the picture and asked if I wanted him take a picture of me with the blaze in the background. I graciously accepted his invasion and he took the picture and then explained "that is a Burn Pit. Most of the waist is discarded that way. KBR is the company that takes care of that." Then I responded with at tone that was clearly asking a question by saying "KBR"? Then Daniel responded "They run the DEFAC, billeting, base transportation, and the laundry and more." I then responded, "I knew they were one of the world's largest suppliers of contracting service but until this moment I was unaware of how deeply rooted they are in this war. But since it is waste and other bad things that they are burning, I am glad it is occurring over there."

Three days after Malcolm returned to the WarWizard shop, I met another leader returning from leave named Marshall. In the pecking order of contractors, Marshall was third in line in the WarWizard shop. He was an African American who was also about forty years old. He was about six feet tall and about 220 pounds. Marshall had the patience of the Biblical figure Job. I admired his leadership skills; He always remained cool and calm regardless of what was thrown his way by unruly subordinates. After getting to know Marshall, I knew that if he could not lead a subordinate down the right road peacefully something was wrong with that worker. Whenever Malcolm met these unreasonable personnel challenges, first he would very calmly repeat what he was asking and almost always the person would indeed get in line with what he is asking. He was very technically sound and a good listener. I viewed Alfred, Malcolm and Marshall as examples of the proper ways to lead men. One would have to be blind to miss the leadership traits that they were exercising. Their practices created a more cohesive work environment.

Days later Alfred replied to my request for a paper copy of my leave request by informing me that headquarters did not render a copy. I immediately thought to myself how unbelievable this place and its processes are. I recalled the first issue with the form was the fact of the leave requestor was not required to sign the form and now being told that I (the requester) will not get a copy of that request. As I stated earlier, I have been an employee of various government and civil employing entities and yet I have never seen this standard of *no copy* for any official action applied in any professional organization. Therefore in order to protect myself later in case no one remembered that the leave was approved, I sent Kimberley an email that questioned if the dates that I had of the approved leave were correct. She responded via email that the dates were the proper dates. Now I had a confirmation and copy of my approved requested dates in spite of their deprived procedures.

Three weeks later my new friend Richard returned to the facility from his twenty five day vacation back in the States. We exchanged dialog about his trip home and happenings around the shop. He asked "why are were so many of the FSRs assigned to Baghdad missing?" I informed Richard that ten of them had been sent to the WarWizard Facility in Kuwait. Richard then informed me that he has been to the Kuwait Facility and their practice was to work the temporary people to death while their personnel relaxed. Based on the description that Richard provided, I had little desire to be a part of the Kuwait WarWizard Team. I expressed to him that I was glad he and his wife made it back safely. Richard then found great pleasure in mentioning to me how he had such a good time driving his BMW.

My training at the shop continued for a few more weeks with Chase continuing with most of my training. Upon conclusion of my training, I was then assigned to a three man install team consisting of Richard, myself and Jameson. Several three man teams were created that consisted of one senior member and two newbie's. Richard was our team's senior member. A different team rotated the work each time a vehicle came in for maintenance.

Both Richard and I were employed by the Primary Contractor, CSS. Jameson worked for a subcontractor of CSS identified as *Mega Electronic King*. As I stated earlier there were numerous other subcontractors from several subcontractor companies as well as DoD employees working in harmony on this contract.

It was a few hours into our work day when I heard the faint sound of an explosion followed by a slight ripple on the ground from the blast in the distance. Shortly afterwards there was a large cloud of continuous black smoke. We were informed that a car bomb had exploded in downtown Baghdad; blowing up a building with over one hundred people killed. Immediately I felt a haze of sadness penetrate my thoughts. Whenever that many innocent people are slaughtered it is a tragic day.

That was a terrible thing to happen but that episode may as well have been one hundred miles away. It affected downtown Baghdad directly but had little effect as to what happened on base with our operations. Incongruously our workday continued as if nothing had occurred being mindful of our mission at hand. Later that night when I talked to my wife on the phone, she informed me that she heard the news report about an explosion in Baghdad that day which concerned her. I explained to her that there was no real reason for any concern for my safety because the explosion happened outside of the base. Understandably my wife's concern was valid from her point of view. That's why I never mentioned to her what I had previously mentioned to my father about hearing machine gun fire in the distance every night. My father was a former Army trooper and I am sure he was somewhat concerned for my safety but understood that there was no reason to mention those type of details to my wife or my mother.

Day by day our work group of three began to exhibit some signs of dysfunction; usually it was the same member. Often when the three of us worked on a vehicle, Jameson was slow finishing, thus holding up completion of the system installation. Jameson's work ethic was beginning to annoy our Floor Lead Malcolm. Just a few days earlier, several times there were vehicles in the bay in need of repair and while everyone was working, Jameson would grab a chair and sit down and work on crossword puzzles. On two other occasions Malcolm had to tell Jameson "If I am working, everyone else needs to be working." Apparently Malcolm's word went on deaf ears because two days later Jameson was once again solving crossword puzzles when there were trucks in the bay in need of maintenance. With anger and frustration in his voice Malcolm confided in me saying "This guy is making me look bad because I am the one that sponsored him to this program. I had no offer of resolution for Malcolm; I guess my role at the moment was to be just a sounding board for him. It is one effort to make an enemy out of those who work against you, but to work on making an enemy out of the person who sponsored you did not make much sense to me. But there was one thing that was clear; Jameson at age 55 still had a lot to learn.

CHAPTER 11

THE TOSS

The three of us continued to work as a team on the installations until Richard started to complain about Jameson's substandard work habits. Jameson continued his practices of walking away from the work or was just working at a slow pace. Therefore, most days Malcolm made a slight change by rotating Jameson on other teams.

Today Richard and I welcomed our newest team member, USAF Retired Bruce Love. Bruce was Pilipino, slender build who appeared to be about two or three years younger than myself. Bruce's knowledge of the equipment became apparent quickly. Bruce fit right in and his work ethic was needed and much appreciated. As a result, again the maintenance team flourished. Richard and I now had nothing to complain about as far as the work was concerned. But wouldn't you know it, some people are never satisfied with harmony and my supposedly good friend Richard is my prime example.

Since Jameson had been fazed off our team, Richard started to complain about both me and Bruce's work. There was no legitimate cause for his complaint; he just wanted to complain

In the meantime, whenever the work was done sometimes we would congregate in the break room and talk to pass the time. It never failed; in just about every break period Richard would always find it necessary to tell us about his BMW. It did not matter if you were a veteran who had heard it all before or newcomer to the facility, you were going to hear about Richard's BMW. It was really comical to witness Jameson play his usual *yes-man* role by saying "Yah Buddy!" in support of Richard's BMW claim to fame. To add insult to injury, Jameson remained a loyal Richard fan in spite of Richard mocking Jameson's work ethic and auto ambitions plans to get his wife a Buick or Cadillac. It seemed that these two were a match made in; well, I am not sure who sanctioned this union.

At times when we were not being entertained in the break room by Richard and Jameson, we tossed around a football for a little relaxation. I had been told that this was a treasured practice that occurred prior to my first day here. It was about 10:30am and the facility had just completed eight tough vehicles. Since the work was over it seemed like a good time for a break so I picked up the football and tossed it to Richard. Richard had slightly shorter than normal arms for his stature but when he threw that football back to me it was a tight spiral with strong velocity. As I was about to return the throw to Richard without saying a word Malcolm motioned to me not to throw the football. The woman who monitors the contract operations had just walked in. She observed Richard toss the football to me. At which point she viewed the event as inappropriate and immediately made an unfavorable report to Headquarters Baghdad and the program manager back in Virginia. She also informed Malcolm of her observation of one of the FSRs reading his Bible. We were all taken back a little by the conclusions and report the inspector had made about our shop. Now mind you, at the point of her inspection all the work in the bay had been completed.

Needless to say, our football tossing was over for a while, and Daniel's Bible reading now could only occur in the break room. Malcolm later informed the shop about our management staff in Virginia's response to the inspector's report. To summarize, they expressed that *"since she witnessed only one toss and then the football" was put away*, to make the conclusion we were playing was premature. Our management team back in Virginia gets nothing but kudos from me for their stand on this issue. Remember, when the contracting officer arrived all the work was done. So when a contracting officer fails to realize that contractors working twelve hours a day in a war zone with no days off should be afforded with some type of stress relieving activity, they have gained a notable mention of being an enemy when there is not a foe to face.

Later that evening Malcolm asked me to pick up a new inbound from the airport. It was about 11:30pm when the newcomer Darrell Ryan's flight touched down. Darrell was an African American male who appeared to be about thirty eight years old. His stood about five feet eleven and about 290 pounds. His stature seemed to be that of the football player Earl Campbell. When I finally made contact with him, he was smiling and grinning from ear to ear as he introduced himself. After introducing ourselves Darrell had one request "Man where do you get food around here at this hour?" I responded, "There is only one DEFAC opened twenty four hours here on the COB." "Well, let's make it happen because I am ready to put some food away" replied Darrell. We headed to the DEFAC and for about two hours we enjoyed some food and conversation.

Another day at the maintenance facility started with another decree from Baghdad's Program Manager Rusty. Now I will admit eating three big meals a day whether you are hungry or not will put on the pounds. Therefore, instead of eating that third meal each day, I decided to jog. Obviously, someone saw me jogging during the evening lunch break and took issue with it. I come to this conclusion because in our subsequent morning meeting the evening lunch policy had changed. Alfred began the meeting by stating, "You are authorized to do whatever you want during the noon meal, but the evening meal must be for dinner only. If you want to go to the gym or anywhere else you cannot do it during the evening meal time because it is not authorized. Therefore, if one does not go to the DEFAC for the evening meal, their only other option is to stay at work.

Alfred continued to explain the new ruling by emphasizing that change is for contractors and does not apply to the DoD civilians. He also reminded everyone of the authorized time for that meal was 45 minutes. I stood there in a state of disbelief because according to my thoughts, once again the managers of this organization were are getting involved in matter that should be left alone. It would be suicide to even think about jogging at lunchtime when temperatures were at its hottest. Once again I find myself yielding to the *"powers that be,"* realizing that I must comply with their demands no matter how irregular they appear to be. After all, they were writing the check. So from that day forward I decided to spend my dinner break eating instead of continuing to work.

CHAPTER 12

A MATTER OF CONSIDERATION

From day one on the job, I found myself starting each day wondering "what's in store for us today?" When I arrived to work on this day today a few of the guys were taking tally as to who could confirm whether or not there was a missile attack on the Camp the last night. I did not hear any sort of attack last night, but the Baghdad installation was so large that an attack could have occurred on the Camp and I never heard it. Hearing talk of an attack usually would cause a little quiver in my stomach, but for some reason I felt no fear or unrest today.

After the morning meeting Richard and I were sent to work at one of the other work sites on the installation. Along our route to the work site, we passed the living quarters of Pilipino Nationals. This area was known as TCN's housing. As we passed these quarters, Richard informed me that about nine months earlier there had been a rocket attack that hit the Pilipino living quarters. That story was certainly believable, but Richard was such a talker, storyteller and sometimes liar, I was not sure what to believe.

Travel time to the work site was about a ten minute commute. By the time we arrived at our destination, Richard had managed to massage my ears once again with his literary rampage about his beloved BMW. In addition to his usual rhetoric about his BMW, the story changed slightly because now that he had found a larger BMW online, he was thinking of trading his in. Richard continued on; he began boasting about his relationship with the program manager Edmond Gibson. Richard thought it was important for me to know that Mr. Gibson was from Mississippi and was is acquainted with his aunts because they went to school together. Richard continued to brag about his tie with Mr. Gibson by saying Mr. Gibson instructed management in Baghdad to limits the number of Temporary Duty Assignments (TDY) that he was required to be a part of. I concluded to myself, "Okay Richard if you say so." Favoritism in that manner due to such an insignificant link seemed a little odd to me, but stranger things have occurred since signing up for this contractual adventure.

As we worked and shared small talk, Richard exercised little discretion as he disclosed some disturbing information to me. It started by me mentioning to Richard that I had met a guy from a previous assignment going to Afghanistan at the Post Exchange with the same last name as he. Richard immediately validated my comment by saying, "Oh yeah, I was in the Army with him! Because we had the same last name people thought we were brothers. He stayed in, but I got out after two and a half years." "That was a short enlistment," I responded. "Bro, they put me out, and before they put me out they were trying to put me in jail," Richard remarked. With a puzzled look on my face I questioned, "Why was the military attempting to put you in jail?"

Richard explained, "Bro, I had this squad leader that just stayed on me. He harassed me if I was 5 seconds late or anything that he thought he could get away with. So one day I caught him by himself and I said why don't you talk some of that Bull now? Either I swung at him or he swung at me, I don't remember, but we were fighting. I hit him on the head and knocked him out. While he was unconscious I taped him up with some duck tape. I taped his arms, legs and his mouth and then I put him in a dumpster." At this point my wrench had stopped turning and I stood motionless in utter disbelief of what I had just heard. Now I knew what my ears had just heard him say so calmly, but my brain did not believe it. Therefore, in an unbelieving tone I stated "Yeah right, you actually did that." "Yes I did throw him in the dumpster!" Richard responded without hesitation. I needed a little more info so I questioned again loudly, "You actually taped a person up and threw them in a dumpster?" "Yeah Bro, if the person operating the garbage truck had not heard him mumbling the next day, he would have been crushed to death" Richard explained. Still not convinced that what I was hearing was based in truth I uttered to Richard, I then questioned "What level of attempted murder did they charge you with at your trial?" "Yes, they had charged me with attempted murder and were trying to get me a ten year sentence" Richard eagerly replied. "Well what happened?" I questioned. Richard continued, "I had a good lawyer that got me a plea bargain of *A General Discharge*. At the two and a half year point I was offered the General Discharge so I took it and exited the military." "You know you were supposed to go to jail, right? Because if that happened the way you say it happened, that was attempted murder," I responded. Then Richard voiced" Bro, I had a good lawyer, and he got me off."

At that moment I had no thoughts; I was numb; I just sat there saying nothing. It was bad enough to perform an idiotic stunt in your youth but to boldly tell someone of your ill thoughts and dreams as an adult seemed strange. After hearing that confession I knew I had one of three categories to place this guy: He was either a liar, had no conscience or just plain obtuse or perhaps it was all three. The only position I could take at this time was this man required the deepest scrutiny. ***"Give them enough time and they will always show you who they are."*** is what my father would always say and this guy was on display. From my prospective, this would be one of those secretes that I would have taken to the grave.

Instead, he wore that episode like a badge of honor. At this point I was starting to get a completely clear view of what I was dealing with when it came to Richard; someone who could not be trusted.

Usually the vehicles that filter through our work-station are assigned to the US Army. But today was one of the rare occasions that this particular Humvee was assigned to the US Air Force with Air Force personnel managing it. The personnel were two airmen, a female SrA and a male A1C, dressed in desert flight suits. I recall only one other occasion when I saw Air Force personnel in desert flight suits. On this occasion I asked them if they would allow me to take a picture of them. They kindly agreed. Being that I am retired Air Force, I cradled a slightly biased sentiment towards the young airmen so I wanted to personally make sure they were taken care of. We engaged in a brief moment of small talk comprised of military related topics then I joined Richard and Jameson to complete the work on their vehicle. I had never installed this third type of jammer in a vehicle before so I questioned Richard as to how it was to be installed. Richard rattled off some instructions to me and I followed them exactly as he conveyed them. As I began to move to my next task, Richard aggressively approached me loudly saying, “You did that wrong!” “What are you talking about?” I questioned. “You were supposed to tie it off the other way” he rudely shouted again. Richard’s rude and elevated tone was witnessed by our awaiting customers, the Airmen. Typically, Richard just had that knack about himself. He always tried to put you down to elevate himself, and I was about tired of it. Therefore, I motioned to Richard, "Richard walk with me for a minute" I voiced. We walked about twenty five yards away from the vehicle. I lit into Richard saying, "Richard you discredited me, but in addition this facility by doing what you just did." "What do you mean?" Richard questioned. "You made it look to the customer as if I did not know what I was doing. But you also make this place look like it has unskilled workers" I responded. Richard defended his actions by stating "All I was saying was it was done the wrong way." "That is not the point, did you not see the customer soaking up every word you were saying" I responded. "Well that's just too bad, you just got to tough it up and stop being so sensitive about everything" Richard responded. "Then you install it." I voiced and walked away. Richard stood there for a moment puzzled but then finalized the system install. From that point I was aware that Richard was no friend of mine. In fact he was a friend to no one. He approached you with a grin, but the only person that he was truly a friend to was himself.

About two weeks after that episode, Malcolm was conducting a class on the systems we installed and how we should conduct ourselves when dealing with customers. During his lecture, Malcolm highlighted a point that he stressed needed to be addressed. Malcolm then summarized "Whenever there is something that needs to be corrected with a co-worker, *never do any correcting in front of the customer because it discredits both the technician and this facility."* After Malcolm made his comments I looked over at Richard with a boisterous beam, but he would only look away, because Richard was one who could never acknowledge when he was in error.

Day in and out set the stage for witnessing numerous skits featuring Richard; so much that I coined a term to describe them, *Richard-ism*. By now nothing surprised me with Richard. It wasn't a matter of wondering if Richard would perform for us, but a matter of when. Case in point, on this particular day, Richard, Jameson and myself went to lunch together and decided to stop by the Post Exchange on the way back from the DEFAC. Richard and I were the only ones who purchased items. We had a few minutes left of our lunch hour; therefore instead of going back to work with items, we stopped by the housing area first to drop off our goods. Richard and I got out of the truck leaving Jameson in the vehicle while we made the pit stop. The trucks engine remained running. Richard's living quarters were only about thirty seconds further from where the truck was parked than mine. I returned to the truck in about two minutes. While we were waiting on Richard, Jameson and I sat in the truck recapping on Malcolm's system installation class. Still waiting on Richard, thirteen minutes had passed when we realized we were late for our return to work. I finally told Jameson, "Richard will just have to walk back because we are already late waiting on him."

We were not leaving Richard out to dry too bad because walking back was only about seven minutes away. When we returned to the work center we were about ten minutes late for our evening system training class. Low and behold when we get to class, there sat Richard at the front of the class appearing calm and relaxed.

After class I asked Richard why did he not return to the truck at the housing area? He informed that he walked back. Puzzled and steaming with disgust, but with reservation I asked him, "Why didn't you tell me you were going to walk back?" "Man you know man I had to man you know I had to get back" Richard replied. I did not say a word. I merely walked over to Malcolm and informed him of what caused me and Jameson to be late. I was still discretely fuming as I further explained to Malcolm the issues I was having with Richard; the biggest one being his lack of consideration for all others. Malcolm listened attentively as I concluded my complaints about how I believe Richard is into self and not a team player and then requested not to work with Richard from this day forward.

Malcolm took immediate action and removed me from Richard's team. Two days after I left the Richard team, Bruce also made the same request and left the Richard team. Finally the only person in the facility that would have anything to do with him was his yes man, Jameson. Though I was thankfully relieved from duty on Richard's team, I was not immune from witnessing

Richard-isms every day. I noticed that whenever Richard and Jameson were slated to work on a vehicle together, to bring attention to himself Richard would stand at the entrance of the break room and yell out at the top of his lungs "Jameson, it's time for work!" And like a true yes man Jameson would come a running. But the sad truth was, as much work as those two avoided, it was a good idea to announce to everyone the few times when they actually worked. That duo wasn't fooling anyone. Everyone in the facility was aware of their antics.

After leaving the clown team of Richard and Jameson I sometimes worked with Darrell Ryan. On one such day he confided in me that if it had not been for some good NCO's he may not have been able to retire, and would have possibly exited the military is shame. According to Darrell, when his group was in formation there was a no striper that was always later for formation. On this particular day he questioned the troop as to why he was late. The troop responded "F**k you;" in front of the entire formation. Darrell at this point tried to squeeze the life out of the soldier. It took several other NCO's to remove his hand from that young troop's neck. At first Darrell was brought up on charges, but they were later dropped, according to Darrell because a lot of the NCO's that were present went to bat for him. At that point, Darrell was advised to retire and he did. I already knew Richard and Jameson were jokes but now questions loomed with Darrell's temperament.

CHAPTER 13

EIGHT DEVILS

Months into discovering Richard's character unfolded and caused me to conclude that as far as I was concerned the name Richard West stood for loser. And perhaps if I ignored the Richard-isms, enough it would go away. Malcolm was aware of the hard line and distance I had put between myself and Richard. As a result, he devised a plan to perhaps get us to at least be cordial to each other. In order to accomplish a means to an end, Malcolm had the three of us meet outside on the shady side of the work area and he put a challenge to us. Malcolm plan started with these words, "Guys do each of you love me as Christian Brothers?" We both answered "Yes." "Then if you love me you will work out the differences you have between each other. So I would like it if you shook hands and made up." Malcolm pleaded. Richard extended his hand toward me, but my hand remained motionless. Then I turned my sentiments to Malcolm stating, "Malcolm, there is a long line of problems associated with this man." Then

Richard voiced, "Kingston, I have changed my ways and I am getting right with God. Not only do I apologize to you but I have also apologized to my wife and kids for all that I have been doing wrong because I have been doing wrong a lot." Now since Richard was speaking from a religious standpoint and asking for forgiveness, two different biblical vantage points entered my head as to how I should respond. On one account the question of *"Can the Ethiopian change his skin, or the leopard his spots?"* was brought to the forefront of my thoughts. Then my thoughts reflected to another vantage point in the Bible. Peter asked *"How often should I forgive someone who sins against me? Jesus answered seventy times seven."* In fact, throughout the Bible there are numerous examples of forgiveness rather than the leopard example. As a matter of fact, with all things considered, I had a person in front of me who appeared to be offering a sincere apology. I felt obligated not refuse. Therefore, I extended my hand, accepted his apology and we agreed our disputes would be forgotten.

For the next few weeks Richard started carrying a Bible and speaking about the Bible and God. You really could see a difference in him. This new walk that Richard was attempting reminded me of some compelling words revealed to me by my Grandmother Creola when I was about nine years of age. I felt an urge to share those words with Richard. Therefore, I pulled Richard aside and warned him of my grandmother's words, *"when you try to walk with God after doing bad for a long time, eight demons will come after you to reclaimed what was theirs."* As Richard listened attentively, I expounded on my Grandmother's words stating, "What was theirs is you, so now is a time for you to really be careful." He thanked me for the counsel and reassured me that he was in this thing for the long haul.

A few days after our conversation, Richard decided that he was going to be baptized again and invited me to come to the baptismal. I wanted to show Richard my continued support therefore I did attend. His work sidekick Jameson was also in attendance. Until this moment it was unknown to me that the minister to perform the baptismal was one of the supply technicians who I usually saw daily at our work facility. He and his supply teammates were responsible for supplying, securing and managing the jamming equipment that we installed. Their supply office was located in a small building adjacent to our maintenance facility. Previously and unrelated to Richard's new walk, Jameson was given the title of Bishop at this same church where the ceremony took place. Thus, Jameson took a special interest and participation in Richard's ceremony. It was a good ceremony and I was actually happy to witness Richard's continued walk in the light.

When I learned of Jameson's appointment of Bishop I was somewhat astonished. My flabbergasted position was because of my observations of his work ethic and interactions at work. Any Minister who would appoint this guy as a bishop forced me question (at least in my mind) his judgment. But I had to consider the fact of the matter could have simply been we were in a war zone and there were just a limited supply of personnel for selection. So I gave them a pass on that matter at that juncture.

Since Richard's transformation, during down time at work the topic of conversation seemed to always end up on something Biblical. Sometimes Jameson seemed to put himself at a higher spiritual plain than perhaps he deserved. I guess he was trying to fill the role of Bishop that had been handed to him. There was a time in particular when Jameson decided to flex his spiritual knowledge muscle with me. He came to me unprepared however because I assume he thought I was not versed on the Bible. There was no way Jameson could have know that as a child growing up my parents forced me and my siblings to go to church even when we sometimes did not want to go. During this time of forced church attendance, I actually paid attention and learned quite a bit of Biblical teaching.

While we were waiting for the next influx of vehicles, Jameson boasted to me that I would not understand the Biblical teaching that he had obtained which goes all the way back to the Garden of Eden. He further claimed, "These teachings are what I am increasing Richard's knowledge of God and the Bible with." I then responded, "Before you can teach at any level, first you must at least be familiar with the material that you confess to have this higher level of understanding of; Right?" I questioned. "I don't understand what you mean" responded Jameson. "What I am saying is, in order to teach on a certain subject, in your case the Bible, you must have an understanding and knowledge of the area of the Book that you are teaching from correct?" I questioned. Jameson's silence led me to believe he was still pondering my statements. So I approached him from a different angle. "Jameson, since you found it necessary to inform me that your knowledge of the Bible goes all the way back to the Garden of Eden, I have a question about that Garden" I stated. "What is your question?" asked Jameson. "When Cain slew Abel and was casted out of the garden, who did Cain leave from the garden with, and who did Cain return to the garden with?" I questioned. "In addition, the person that Cain returned with, where did they come from?" I again questioned. There was complete silence from him. "Is there a problem?" I calmly questioned. "I don't know" said Jameson. I relieved Jameson's baffled mind by answering, "Cain did not leave with anyone, but he returned with a wife. The Biblical text does not say where she came from. Jameson offered no response and that ended our Bible lessons for that day. I cautioned Jameson once again about knowing a person's knowledge base before disclaiming that person. As a result, there were no harsh feelings to mend, in spite of Richard strongly laughing at his religious mentor who had once again been silenced.

It was an extremely busy early mid September day at the facility. I was just finishing up a project I had been working on when I looked up and saw a tall, light skinned gentleman walk into the hanger. His hair was balding on top, appeared to be about 6 feet 1 inch tall and 210 pounds. He came in and leaned against the wall and stared for about an hour. He didn't say anything to anyone; he just stood there with a disturbed stare. I didn't spend a lot of time wondering who he was; but I presumed he was an inspector. Finally a female from the supply office addressed us, "Why won't someone say something to Bailey?" "Is that his name?" asked one of the FSRs. "Yes, he is a new hire" she replied as she left. We all thought it strange that no one forewarned us of Bailey's arrival or introduced him on his first day. Bailey was an interesting fellow. Usually you would take it upon yourself to do the introductions if no one else initiated it. Nevertheless, we all welcomed him then returned to our busy maintenance projects. Bailey did not seem to have the most diverse background insofar as maintenance or electronics, but when it came time to work he gave you 100 percent of whatever he had. Darrell Ryan left for Camp Taji a couple of days after Bailey Arrived. He emailed me a couple of time and informed that Taji was great and I should volunteer for it, but I never did.

Richard stayed on his righteous course a few weeks longer but as a wise man once said, *"All good things must come to an end;" and* Richard was no exception.

Remaining a positive Godly person seemed too much for him. With each passing day, glimmers of the old Richard seemed to surface.

On any given day at the job you never knew what would come your way; and today was no exception. The engineers out of Fort Mammoth, New Jersey could not figure out where to place the antenna for our equipment on the Striker Assault vehicle. They would pass down modifications that were constantly being changed. Directive after directive was introduced but nothing seemed to satisfy the problem. Finally, one of the shop's DoD guys (white male, whose first name was the same as mine) was tasked to reposition the antenna. Through previous interactions with this particular FSR, I discovered that he was inexperienced and not too mechanically inclined, but he appeared to be a decent guy. Therefore, I volunteered to assist him with his assigned task. We mounted the antenna and the cables three different times. On the third time we were given the green light from Marshall that this mount was a done deal. While we were in the middle of wrapping up the installation, Richard came over to us insisting that our installation was not being done right. Annoyed by Richard's comments, I responded "we got this Richard; Peace." Richard walked away for a seconds then came back instructing us that we needed to reroute the cables. I again said "Peace brother." The vehicle was about to be released when Richard solicited Marshall and one of the engineers to get the cables repositioned. Since Richard was relentless about the cables he managed to persuade the engineer to change his mind on the correct method of installation, therefore Marshall had no choice but to tell us to reroute the cables. This was a fourth change, but the change really accomplished nothing. Richard walked by and gave us a devilish smirk. It was clear that Richard was basking with content since his sole interference caused us to have reposition the cables and antennas. Once again Richard had to forced his will and forced us to make the change. I do not think for a second he was attempting to do what was right or noble or what was required; he was just trying to have his way. It appeared as if the eight devils my grandmother spoke of had returned to him. We not only had our old Richard back, this friend was now much worse.

CHAPTER 14

CAMP MEADE

It was the end of a hot work day in mid September when Alfred announced in the evening meeting there was a Fly Away requirement for Camp Meade. A Fly Away refers to personnel being sent to another COB or FOB's location to provide technical support. The majority of the time the mode of transportation for an outside the wire venture for contractors was to Fly Away to a site by helicopter. At the time of the announcement, Alfred admitted he did not have all the details, but he was sure the assignment would occur in about a week requiring two FSR volunteers to complete.

In addition, Alfred stated with a cheerful voice that if he did not get any volunteers two persons would be selected and would be *volun-told*. At that instance some of the words of my dad and Uncle Billy echoed in my mind; when it comes to military type endeavors. "*NEVER VOLUNTEER FOR ANYTHING*. Even though I was tempted to volunteer for this mission because I could use a break from COB Baghdad, that old advice prevailed. A few days had passed without anyone volunteering, so Alfred announced at the end-of-day meeting that Jimmy and I were volun-told. In a way I was glad that I was temporarily getting a break from the Baghdad madness, but at the same time I had questions regarding the safety factor of where I would be going. I tried to research the internet for information about Camp Meade but found very little. Alfred later informed me that Camp Meade was a few miles outside of the Baghdad Metro Area and the travel time would take about ten minutes by helicopter.

Another reservation I had about the trip was the partner that was selected to go on the mission with me. I did not know Jimmy very well. What I did know or I thought I knew about him was that it appeared he could not be trusted. My reason for taking this position stem from a previous event that occurred between us. On one occasion I was working on the top of a vehicle installing a rubber seal that water proofed the top of the vehicle. Jimmy came out, looked at what I was doing and quickly went in the direction of the commander's office. Within sixty seconds the commander was at the back of my vehicle angrily questioning me, "How can you protect a soldier by installing that equipment in that fashion?" I was blindsided by his approach and action, but I quickly reevaluated my method and discovered I was indeed installing the seal improperly. I responded, "It will not protect in this fashion, I will make the necessary changes." And what could he say after that; Nothing that's what. I sort of took the angry air out of his sail. He left and I never heard another word of it. Besides, I was a contractor, my paycheck was a lot larger than his and there was no way I was going to let that level of income be jeopardized. After the incident my thoughts reverted back to Jimmy and his actions of checking out my work and then heading to the commander's office. "What a piece of work" I thought to myself. He could have very easily informed me that I was not installing the seal properly. But instead, he went to the commander and placed me in a negative light. At a minimum this was not the guy I wanted to be in the same room with let alone serve with on a two-man mission.

I can't say I was looking forward to the mission because of the reasons just stated, but I won't say I dreaded it either. I guess I just felt a slight hesitation due to the fear of the unknown. I needed a dose of speaking with my wife so I contacted her that evening on the phone and informed her of my selection for the Camp Meade Fly Away. I didn't like the notion that I possibly would be unable to verbally communicate with my wife therefore I wanted her to pay attention to her e-mail just in case phone contact was not possible. As usual, by the time we concluded our call that evening I was recharged.

I stayed in contact with Alfred as to any additional info that he might have concerning the Fly Away. The first update was that we would be leaving the upcoming

Sunday and we would only be performing maintenance on the Green Box Jammers. The two most common types of vehicle jammers we installed green box and tan box jammers. The third type jammer was being phased out. Our mission at Camp Meade required us to work with the Green Box Jamming systems. Both performed well but they were two different types of systems. Each system type require its own different test equipment to ensure that system functioned properly. This Camp Meade mission required Jimmy and I to make adjustments inside the green box systems that upgraded the electronic frequencies used.

It was Friday when Alfred informed Jimmy and I that we would be leaving Saturday instead of Sunday. Since the departure date moved up I had not started to pack yet. Therefore, I was released at noon to pack and take care of any loose ends that needed to be done prior to leaving. Traveling light was the name of the game with these types of Fly-Aways. Therefore, with the exception of our personal tool bag all equipment needed on this mission was sent ahead of our departure. When I phoned my wife that evening I informed her of my departure day and again confessed my love to her. Alfred had not given us a return time frame so I reminded her to stay tuned to her email.

Our show time was 0600 at the COB Victory Helo Launch Pad. Alfred drove Jimmy and me to the launch pad in preparation of our departure. We attempted to check for our reservations but the flight officials could not find any flight record of us. We were then sent to the Liberty Launch Pad. Again the officials found no record of our flight but they were able to tell us the time of the next flight; which was 0600. This was the time of our scheduled flight so we assumed this was our intended flight, so we waited. While the three of us waited for the helicopter, Alfred and I engaged to a lengthy conversation about our past electronic and maintenance experience. To our surprise we both had the same type of aircraft maintenance background except, his was with the Navy and mine with the Air Force.

The helicopters arrived right on time. There still was no record of our passage, but we were allowed to board because space was available. The flight attire requirement was earplugs, long sleeve shirts, long slacks, eye protection and closed-toe shoes. The body armor and helmet were also a requirement and anyone without that protection would not be allowed to fly. This was my first helicopter flight so as for me; this upcoming flight was quite exciting. We loaded our bags and in seconds we were airborne. As soon as the helicopter peaked above the thick brick walls of the compound, the enormous size of the city of Baghdad was in full view. At that moment a stark reality was brought to the forefront of my mind. There were gunners on each side of the helicopter. I had seen gunners on helicopters many times from the ground before, but when you are aboard the aircraft that requires gunners for protection the distant war suddenly get close and very personal. However, since this was a once-in-a-lifetime experience, I took out my camera and started taking pictures of Baghdad and its landscape. The flight took about 15 minutes. Just as the city started to end and the green marsh farmland began a small military compound appeared. And with a blink of an eye we descended upon it. Jimmy and I departed the helicopter and in about 25 seconds it was again airborne and quickly disappeared.

Jimmy and I located the command post and attempted to meet with our point of contact, Staff Sergeant Bob Fickland. No one at the Command Post had any idea where Bob was therefore a runner was sent to locate him. I wondered who was in charge of organizing this mission we were volunteered for. Apparently our flight was never scheduled, we just happened to get transportation because space was available. Then we got to this destination and our point of contact is nowhere to be found. I hoped these incidents were not going to set the tone for the rest of our time there. In the meantime we had no other option but to wait. And wait we did. During this time I ventured to the portable potty. To my surprise, this trip the latrine was as about as far as one could get from the running water that operated the toilets at Baghdad. There was nothing but the ground to capture my urine. I figured there must be some sort of mistake. Perhaps this was an old unit that was no longer in use. No big deal, I thought, therefore I went to another potty only to find the same setup. By this time however, nature forced the issue, I had no choice but to go then and there. I laughed to myself thinking, "So glad nature wasn't calling for number two."

Finally, about 40 minutes later the runner came back with the news of Staff Sergeant Fickland's location. Bob was in church. No one at the company post knew of our mission therefore they could only give us shelter from the heat until Bobs arrival. The command post persons gave us water and offered us a location that was indoors, but Jimmy and I elected to stay outside in the shade. Jimmy and I did not have much to say to each other. As far as me, I was not about to take my guard down with the guy that had witnessed me making a mistake and did not tell me, but rather told on me. I could not speak for Jimmy because I'm not sure what he had against me. Though we never exchanged harsh words or anything, we never had much interaction between us. All things aside we were there to assist in saving lives. And that purpose was to give the soldiers the best possible chance to get back to their loved ones with everything in tack.

Finally, a Specialist at the E4 level arrived and introduced himself and gave information as to what we were to do. Then we followed him to our lodging tent. He also informed that they did not expect us until tomorrow. So somewhere between Alfred and their management the timeline got crossed because I remembered our original time to depart was moved up one day. This was a very small FOB. The Specialist then gave us a quick familiarization tour of the area. There were phone and internet services located within the Recreation Center. There were phone but not DSN access phones. Therefore, for that time my communications with my wife would be solely, e-mail. There was a one day laundry service and a small, but good dining facility. There was also a Hagi Bar, restaurant and a small post Exchange ran by local Iraqi's. The Camp Meade's dimensions were only about one square mile. We were showed where we would work on the vehicles and asked when we wanted to get started. Jimmy and I agreed to get started at 4:00pm that evening; after we got about 90 minutes of rest.

When we finally met Staff Sergeant Fickland that evening he was extremely glad to see us. He quickly informed us that we had his full support. We were briefed on all there was to do at his location and worked for only about 2 hours that day. After that we had lunch and got ready for the next day. At lunch the three of us decided it would be best to start work each day at 6pm after the days temperature started to decline.

The bathroom issue again came up as before. There was still nothing but bare ground under you when you went. This times bathroom exchange I had more than urine to relinquish and I knew there had to be something that I was missing because I knew dropping human waste on the ground could not be the procedure. Therefore, I question the nearest guy that I saw on the matter and he showed me the stack of plastic packets next door used to line the toilet seats.

Basically the packet was placed over the seat and collected the waist. When one is finished, they stand and place the lining into a chemically treated zip locked bag at the bottom of the sanitized apparatus. No smell or any hint of a smell. The chemicals in the disposable kit took care of that. It was then thrown in a garbage can. A detail picked up the can at the end of each day and based on their schedule it would be added to the burn pit. No dirty water or stench left behind as with a *Port-A-John*. Unlike Baghdad, this was a small FOB and I wondered if it was truly safe to burn filthy waste such as this in close proximity of personnel. KBR was in charge of the burn pit actions.

This was an easy duty. Jimmy and I had about seventy vehicles to upgrade in a five day period. Simple math showed that fifteen trucks a day would take care of the total number of trucks. This was a small fraction of the numbers we turned out each day in Baghdad. After Jimmy finished a vehicle I inspected it and he did the same for me; just as Alfred had directed us to do. We worked only about 6 hours each day with plenty of down time mixed in because some of the units either did not arrive, or were late.

Everything went smooth on the job, but as always some type of dilemma would occur. My wedding band was missing. I initially started a search with the trucks that I had performed maintenance on but had no luck locating it. Then I searched my tent and luggage. Finally I told Bob of my problem and he told me a search was underway.

Downtime was something that was new to me since we normally worked 12 hours on and 12 off back at our Baghdad COB. During our downtime we either took naps or hung out in the Recreation Center and watched movies; I also continued my wedding band search. This went on for 4 days and on the 5th day we were off the entire day. Our flight was not due to arrive until the next day so between the movie at the recreation center or resting in your tent, there was no other recreational activities to partake in.

While I explored the camp during some of our downtime, I stumbled upon the burn pit. If I had known what was being burned which was where human and other waste I would have stayed far away from it. I had seen the pit in the distance at Baghdad, but because Camp Meade was such a small FOB it was visible from just about any point on the installation.

Staff Sergeant Flickland made our flight arrangement and we were at the Helo Pad about an hour before the scheduled arrival. The helicopters appeared so suddenly that they seemed to appear from nowhere. There were four other contractors scheduled for that same flight. They all were on the flight manifest (roster) but Jimmy and I were not. The air crew commander was asked by Bob if we could proceed back to BIAP with them and we were allowed to return to Baghdad on that flight.

I enjoyed the stay at Camp Meade. There was more of a feeling of direct accomplishment than that which occurred in Baghdad. I draw this conclusion because for a few days we lived with the soldiers that we provided the life saving aid that helped them get back to their love ones. They were extremely appreciative of the service we provided and often told us this directly. Meade gave me the long break that I needed that would have never occurred at Baghdad. The only rough spots I encountered at Meade were the bathroom facilities; not the bathroom, but rather the distance (about ninety yards) that it took to walk in the middle of the night to get to it. After the experience with Camp Meade's there just were not enough TDYs around for me to Volunteer. Jimmy and I started out as enemies, but for that stretch we had no one else to rely on but each other. He held up his end and so did I. We were forced to trust each other; and from that day forward, Jimmy was no longer an enemy. And as far my wedding ring, Bob gave it to me just before we left. He said one of his troops found it while cleaning and inspecting their vehicle. I tried to give a cash reward for him to pass on to her for what had occurred, but he told me that she told him to tell me that she was the girl from San Antonio and was glad to be able help the maintenance team that helped keep them safe.

CHAPTER 15

LIFES NOT ALWAYS FAIR

The crew started to pile into the truck for lunch. Bailey was not aware of the lunch time routine since it was just a few days after his arrival. So when he finally realized that the truck was about to leave he ran to the vacant seat side of the truck's back passenger door. We thought he was going to get in, but he kept running around the truck for three times, then finally back to the original door and entered the vehicle. I was sitting in the rear of the vehicle and clearly not occupying the vacant seat. I asked Bailey what was the meaning of the ritual he just performed. His response was "I was doing my combat entrance into the vehicle." Two days later the same lunch routine of everyone piling into the vehicles was occurring. This time Bailey ran out to the truck's hood on the passenger side and rolled across the hood consequently landing on the other side. Once again, when he got into the vehicle he was asked what was he doing. He explained that he was doing a combat roll before lunch. On yet another day during the lunch hour, our truck was on its one and a half mile journey to the dining facility. We ventured about half way when I noticed Bailey running alongside the road. We asked him if he wanted a ride; but instead of saying yes or no thanks, his "response was "look…………, a truck………..,is coming." Bailey continued to run in the 135 degree weather and hardly broke a sweat. The blue sky did not render any protection because it did not harvest a single cloud that day. As you might have obtained by now Bailey was in excellent physical shape; but his fitness in other areas was now very much in question.

The next morning Malcolm called me over to his desk and voiced " Chancey I need you be in charge of a small team and make a run to the other end of the base. Now the guys that I need to send with you are Jameson and Bailey." I quickly voiced, "Malcolm I am in as long as it's within the wire." Malcolm continued, "I figured you would be. Do you know the way to the Armadillo Yard"? The Armadillo Yard was one of several storage yards on the base that stored various types of military vehicles; but in particular this yard stored tanks. I replied, "yes, I sure do." "Good, because I was almost forced to send Jameson and Bailey by themselves and I really don't want to do that. I need you to head up a team with them to perform five de-installs at that yard." Jameson and Bailey were informed of the mission and after we gathered all the necessary tools for the project we headed out. The drive was about twenty to twenty five minutes away. The Armadillo Yard was not the ideal location for work that required long hours to complete because there was no cover to work under or any sort of protection from the heat. When we arrived at the location it was about 8:45am. The temperature had already spiked over one hundred degrees, so getting in and starting quickly to facilitate getting out of this mission quickly was imperative. Therefore upon arrival we immediately began removing our jammers from the Bradley Tanks.

All of a sudden Bailey remarked with frustration in his voice, "This is boring." "Well it is what it is Bailey, but don't take light of what we do; we support this mission quite a bit by doing what we do" I responded. "I want to do more than this; I want to do something exciting," Bailey continued. "Like what?" Jameson questioned. "I want to go outside of the wire" Bailey boastfully responded. Jameson and I simultaneously caught each other's eyes in wonderment and a puzzled look on our faces. "Bailey don't you know that if you go outside the wire you may be shot at or possibly killed?" I questioned. Bailey enthusiastically replied, "That's okay, at least I will be doing something more exciting than this, I am just bored." Then Jameson yelled "Are you crazy?" There was no reply from Bailey after Jameson's inquiry, just silence.

Later in the afternoon most of the trucks had been worked on and most of the FSR hung out in the break room. Bailey decided to make a call to his realtor on his cell phone. Being that it was already extremely loud in the break room, one would think that Bailey wouldn't try to establish an important call amongst such clatter, but he did. "I am trying to talk on the phone," Bailey shouted. "This is the break room Bailey!" shouted one of the FSRs. "We can talk as loud as we want in here" added two other FSRs. He then decided to stick his head under a medal table that was set up as the coffee and condiments table. The table was draped with a tablecloth. This was a laughable picture perfect moment, Bailey's head stuck under the table with his rear sticking out from under the table. This priceless moment called for nothing less than a snap shot; therefore several FSR's decided to take advantage of this Kodak moment.

Yes, Bailey had his issues but he loved his country and was loyal and dedicated to the cause. He took our mission to support the troops to heart. I've always said *you should never judge a book by its cover, let the pages unveil its substance*. After learning about some of the chapters in Bailey's life I've concluded that he has been dealt some atrocities that probably have a direct impact on his actions. He revealed that at an early age his dad was blown up in a car. Soon after his father's death, his mother either died or was killed under unusual conditions; he says it appeared to be suicide. No family member came forward to take him in therefore he was sent to an orphanage and later a homeless shelter. He was odd, but when you got to know him you could not help but feel that early on he did not get a fair shot at life. Bailey was a little odd; and yes he did run during extremely hot temperatures, performed a combat roll ever now and then at lunch time, and desired to work outside the wire. Odd yes, but when it was time to work he worked; and always gave one hundred percent to the task at hand. He was simply attempting to do the best job possible, in spite of all of the fights he appeared to face from within.

CHAPTER 16

THE A TEAM

It was now the last quarter of September and I had been in the program for about three months. Ramadan had just begun. Ramadan is an Islamic Fast that is observed daily between sunrise and sunset. It occurs once a year for about twenty five days. The rules of this Islamic religious holiday are simple; if anyone is caught eating during the daylight non eating period they will be arrested. However, unless you were to go outside the confines of the base, Ramadan had minor significance because inside the walls of the base our mission carried on as usual.

Around this same time, about fifteen new personnel arrived for training. I soon discovered that four of these contractors, Perry, Lenard, Bruster (Tex) and Jose Garza added a lot of positive energy to a sometimes flat shop. Although they all had maintenance backgrounds, they had not worked on the jamming systems in which we maintained at this site. Perry was a thirty-eight year old single African American male about my height (five feet and nine inches) but was maybe ten pounds heavier than I. Perry Odom was from Charles City Iowa and had served in the Navy then later the Air Force for a total of ten years. Lenard was also an African American male, thirty six, and married with two small children. He served ten years in the Navy. Lenard had been to Baghdad before on another contract and knew all the hot spots in Baghdad for entertainment. Bruster was a Caucasian male, married, and forty eight years old. He was a true to the heart cowboy who lived in Texas somewhere around Waco; which is why we all called him Tex. Tex was a retired Army SFC with experience in aviation. Jose was Mexican, single, and thirty six years old. Jose's background consisted of varied maintenance contracts he had worked the past few years. Jose lived in Plano Texas. These guys knew each other from previous contracts. Since their arrival, the work place had become an entertainment Mecca. Whether we were working or not, they frequently cracked the kind of jokes on each other that would have made enemies of some people, but they all knew it was always in fun.

Someone who didn't join in on this new light hearted fare in the shop was Richard. On their first day, on two different occasions, Lenard and Perry asked Richard for some simple instructions on installing different items on the vehicles. I don't know if Richard was having a bad day or he was just being Richard but he would not utter a word or lift a finger to help them.

Therefore, I stepped in and walked them through the initial maintenance steps that they both had earlier requested from Richard. Now I wouldn't be completely honest if I did not mention that during the instruction session with Perry and Lenard, I informed them of Richard being the shop loud mouth who usually did not help anyone unless it benefited him.

With the arrival of this new list of personnel came an order to de-install six hundred systems on an array of vehicles because a guard unit from Arkansas was being reassigned back to the states. This order required manpower from the WarWizard Shop to go to the location of the vehicles instead of all six hundred being moved to our location. Besides, the normal scheduled appointed vehicles and walk-ins still occurred. The arrival of such a large number of extra vehicles a week would clog up the normal operations.

To accomplish this task, the shop personnel were divided in half. Seventeen guys were on the newly formed install team and seventeen stayed in the shop. Our Floor Lead, Malcolm headed up the mobile team and our Assistant Floor Lead, Marshall remained at the shop to manage it. I was assigned to the mobile team which included Lenard, Perry, Tex, Jose, Richard, Jameson and others. Later after pleading with Malcolm that he needed some excitement, Bailey was also added.

The six hundred vehicles were Strikers that were located within the wire about 15 miles away from the WarWizard Facility. A tent like structure was erected to conceal our work activities from the sun. Developing a plan to execute the vehicle installs in the most efficient way started out a little sluggish but once we were clear on what was being asked, the process went like clockwork. That is, at least most of us hit it that way. Leave it to Richard to always find a way to do as little work as possible. He spent a lot of time sitting in a folding chair complaining about how much his knees ached. His side kick, Jameson, was no better. At the time Jameson was at least sixty pounds over weight and like Richard who was forty pounds over had trouble climbing up and down the vehicles. Therefore, on our third day of vehicle installs, Richard and Jameson were sent back to the shop and swapped out for two new arrivals. With a fully operational team of workers, one week later the project was over.

As I mentioned before Lenard was a repeat contractor to the COB and he knew Camp Victory like the back of his hand. Consequently, the first Wednesday night after the massive de-install venture, Lenard, Perry and I went to international dance night. On this particular night there wasn't much dancing going on, but we did stay to engage in some of the general festivities. It wasn't like there were numerous other activities to choose from. However, two nights later we made another attempt on Latin Dance Night. This was a jumping hot spot. There were lots of Latin girls and guys and a lot of dancing. Clearly, Lenard was the dancer out of the three of us and he showed no hesitation in dancing with several of the girls. His moves confirmed he had done this style of dancing before. Perry and I were a little reserved at first, but we both found it hard to resist at least trying. I finally did dance a little, but a little was all I could muster being that I had not been dancing in ages and I was not at all familiar with that style of dancing. In spite of my lack of Latin dancing moves, the night was still a refreshing change from the usual routine. But the experience caused me to consider learning that style of dancing with my wife when I returned to the States.

The remote project went well. We spent the next week back at the work shop operating as usual. By the beginning of the following week, the work shop was assigned another remote project. Just as he had done previously, Malcolm split the shop into two teams. This time, I volunteered for the mobility maintenance team. In no time the remote team was in place and operating in a cohesive flow. However, one day our cohesive flow had a hiccup. It was day three of our remote project and before we knew it, it was lunch time. As usual, our same lunch crew jumped into the pickup truck and we headed to the dining facility for lunch. After lunch we went to the Post Exchange. We all agreed to be back at the pick-up by 12:10pm which would give us plenty of time to be back at the work site. We all arrived within the agreed timeframe, that is, all except Bailey.

Perry and I went back into the Post Exchange to try and find him. We also searched for him in the Bazaar store next door; but still no Bailey. We waited for about twenty minutes then I suggested that the bulk of the crew go back and I would stay behind and continue to wait for Bailey to wonder out of the store. The group left with Tex agreeing to return to pick up Bailey and me. About 15 minutes later Tex returned. The first words out of Tex mouth were, "You are not going to believe what that F**k*r did! He was right there at the work site waiting on us when we returned. He ran back." This was the first time I had been involved in a Bailey *run off episodes*, but I soon found out that this was not the first. A few days after our temporary duty ended, Malcolm finally got wind of Bailey's *run off* episode. Malcolm firmly counseled Bailey saying, "Bailey this is the third time that you have run off and told no one in the group you were doing so. So I am telling you now, if it happens again there will be very serious consequences." Being the strong leader that he was, Malcolm clearly got his point across to Bailey because we never experienced this type of nonsense from Bailey again.

CHAPTER 17

The Enemy Appeared

It was now only four days left in September and the NFL had just started its 2009 Season. One day at the shop, Alfred was going around polling the crew as to who wanted to partake in this year's football pool. I agreed to participate and in turn Alfred promised to provide the teams listed for the pool. I was a little cautious about my participation in the pool because previously I heard there was supposed to be a no gambling in the country rule. Therefore, I asked Richard to give me his take on the football pools carried out in the previous years. Apparently I had tapped into Richard's softer side; he had lots to say about the pool. He informed that pools are done all the time not only in our shop, but several shops all over the COB, with a prize of three to four thousand dollars sometimes. He didn't really have a definitive answer about the *no gambling* comment that I had heard. Richard went on to boast about how well he did at picking the right teams in some of the past Iraq Football Pools. He went so far as to tell me that he would take the prize this year because I had no chance of winning with my pipe dream New Orleans picks. As I stood there listening to all his rhetoric, I could not help asking myself "Why did I ask him anything?" Then Richard started his dissertation on the Buffalo Bills expressing how Buffalo would be the team to beat this year with the addition of Terrell Owens and how New Orleans had no chance this year against his Bills. It was comical yet in a small way a bit refreshing to see Richard light up when talking about his Bills. Honestly, I'm the same way about my Saints. With that being said though, I had to put a stop to Richard's rampage. So I blurted out, "Richard you are the only one here who is glad to see a bill. Everyone else pay bills off and gets them out of the way as quick as possible; you know, gas, water, electric, etc." Everyone in the break room got a good laugh on Richard and his Bills that day.

The cutoff for the football pool was finally upon us. The cost was twenty dollars for the participation fee. I made my selections based on who was good the previous year. Since there was still an unclear notion in my mind about the football pool and gambling, I put an alias name on the entry form. And since I was USAF Retried, I called myself "*Air Force 1*." On the day of the kickoff of the NFL season, the shop was released for the first time one hour early by the Acting Facility Commander; DoD Supervisor, Phil. The entire time I had been there, I had never seen the shop released early; even when there was no work in sight we always had to stay. Needless to say, I had no problem with that decision.

New Orleans started out its season with a victory against the Lions at home. I had only missed two games picks that weekend in the football pool and when we came back to work that Monday morning I was the only person in our shop who was still alive in the base wide pool with a chance to win. The Monday Night Football game was the only remaining game to be played for the tiebreaker. Since we all picked San Diego to win, there would be a three way tie if San Diego won. The winner then would be determined by the person who picked the closest number to the point total of both teams from Monday Night Football Game. Well, San Diego did win and there was a total of 44 points scored between the two teams. I picked a 47 point total and the other two entries picked 51 and 49. That was a cool way to start off the NFL Season, a purse of eighteen dollars. At the end of the day Alfred walked up to me and discretely gave me a tightly rolled stack of twenties and hundreds that totaled the purse. Now unparallel to Richard's Bills, I was glad to see those Bills (that is, the dollar bills).

However, when Alfred gave me the money he threw in a curb ball to me. He casually tapped my shoulder and said to me smiling, "It is our tradition that if the winner comes from this shop he funds the next Bar-B-Q." This had never been mentioned, but after such a large winning amount I agreed to be the money guy. Turned that out the next Bar-B-Q was in a few days so I went to the Post Exchange and purchased about fifty dollars worth of meat. As for myself, I needed a microwave oven and those winnings made that purchase appear just a little sooner.

The football pool was a welcomed diversion from the masking reality of being in Iraq. A little bit of excitement was inserted into the day each time any one of us engaged in the trash talking or boasting about our respected teams. I especially was inclined to report (specifically to Richard) any time my Saints came out on top. That being said, New Orleans went on to defeat the Philadelphia Eagles in week 2 and their next opponent would be against Richard's Buffalo Bills. The game was intense at first because Buffalo lead 7 to 3 in the first half of the game with a scoreless third quarter by both squads. However, there are 4 quarters of football and the 17 point final quarter by the Saints made the final score: New Orleans 27 and Bills 7. The outcome of the game had me grinning from cheek to cheek and I couldn't wait until the next day to taunt Richard. The next day as we gathered around waiting for the morning meeting to begin, Richard had all sort of excuses insofar as why his team lost. He was chattering to whoever would listen, "the Bills owned New Orleans for three quarters." I quickly squashed his rationale by responding "Too bad there are four quarters in a football game." "You are an idiot" Richard blurted angrily. "How about that; The shops court jester is attempting to give someone else his title" I responded. He could dish it out but could not take it therefore. The reality of his team losing to mine there was no additional comments that he was willing the exchange with me. The friendly spirit of the games was quickly vanishing for Richard.

The first day of October was finally upon us. I had been jogging with Stan the past couple of weeks in the mornings before work. Stan was a thirty year old DoD guy from Vermont. In order to jog in the mornings with him and work a shower in afterwards required me to get up ninety minutes earlier than usual. This schedule coupled with twelve hour work days was beginning to wear on me. Staying up late and getting up that early was just too brutal on my forty nine year old body. For me it was simple, after work was the only time I would be willing to jog. Stan and I met at my room after work and we briefly discussed an alternative to getting up so early during the morning jog. We were continuing our discussion when suddenly there was a loud explosion, followed by a second and a third. Instantly a flood of people appeared from inside their quarter's doors to see if this was an Army drill or an enemy attack. As we viewed the skies we could see the illumination of anti rocket flairs and interceptors. It didn't take anyone long to realize this was indeed an enemy rocket attack. Within seconds for my first time I sought shelter in a bunker which was only about twenty five yards away. There was no system that warned us prior to the explosions. When I entered the shelter it was dark, but I heard a familiar voice; and I thought even in a rocket attack I cannot seem to escape this guy, that's right you guessed it. Richard was already there with his tong in full gear telling others who sought refuge about the attack that hit the camp about nine months earlier that killed some Pilipino TCN workers. Within minutes after the rocket attacks, two attack helicopters were airborne and in hot pursuit of the enemy. Once things settled down and we were back in our lodging rooms, Alfred came around to the rooms of his personnel to ensure that everyone was accounted for. No one lost their life at our camp that day, nor did we suffer any real damage to the area. Reflections of the weeks past brought to mind the satisfaction I felt winning the football pool victory at the close of September. The victory was highly suppressed by the reality of nearby explosions experienced for the first time in my life this early October day.

CHAPTER 18

OCTOBER MADNESS

The forth of October brought on the 3rd opponent for my beloved Saints that season. The Saints faced the New York Jets in the Louisiana Superdome. One of the permanent members of the Baghdad team was a twenty five year old African American Male from New York whose name was Jojo. Jojo had been in Baghdad for over a year and knew the FSR job well, but his sports knowledge was still in question. After hearing Richard's comments Jojo decided to join in on the Saints bashing. Jojo mentioned "All the New York teams are better than the Saints, they just got lucky last week against the Bills and have no chance against the Jets this upcoming week. He further stated the Giants without a doubt would smash the Saints later after the Jets were done with them.

The game would be aired at midnight Baghdad time. Chase, who was the night shift supervisor, informed me he would be watching our guys knock out the Jets.

Therefore we agreed to watch the game together. The final score of the game was; New Orleans 24, Jets 10. The next morning I was happy to remind Jojo that for two weeks in a row his New York teams had fallen to my guys. "They are just lucky, just damn lucky and there is no way that they will beat the Giants" Jojo insisted. My response to Jojo was with confidence, "We will see."

The events that started off the month of October made me more eager to start my vacation leave set for November; I checked my email daily for my confirmation. Back in July during my in processing in Baghdad with Kimberley, she informed me that vacation leave projections must be finalized by the first of the month in the month prior to your leave. She also explained there was a simple rationale for this practice; airline tickets were purchased in bulk and in advance. Therefore, since I planned to start my vacation leave in November, to stay within the guidelines Kimberley outlined for me my firm leave plans had to be submitted no later than 1 October. I originally submitted plans to fly from Iraq, to Kuwait with a final destination in Texas. However, I later emailed my definite vacation plans of flying from Iraq to Kuwait with a final destination of Alabama to Kimberley on 20 September. This was ten days prior to the specified cutoff date. I received my leave itinerary from Kimberley on 5 October. To my dismay, the itinerary did not have the changes I emailed to her back on 20 September.

Therefore I immediately emailed Kimberley and informed her that there was a slight error in my itinerary which outlined my flight as flying to Texas instead of Alabama. Ten minutes later I received a call from Kimberley. In a stern voice she inquired, "Kingston, what is this about your flight plans not being correct?" I responded, "You did not make the changes I submitted for my flight to depart Kuwait for arrival in Alabama." "You were supposed to notify me of any changes by 1 October," Kimberley voiced. "I emailed you of the changes," I responded. "Well I never got it," Kimberley answered. "Would you like me to forward you a copy of the email I sent you on 20 September?" I questioned. Complete silence permeated the airways for a good forty-five seconds as I listened to the sounds of tapping of keys on a keyboard. I deduced that she was probably attempting to find the email. Finally, Kimberley broke the silence stating, "I will call you back." She called me back about twenty minutes later, "Kingston, I am going to make the changes this time, but next time I won't be able to make any changes." "Thanks Kimberley," I voiced. "No problem," she replied. Again, I remained cordial with Kimberley because even though I was completely right in this instance, I realized that she holds the power. I didn't want anyone or anything to stand in my way of flying out of here. Methods of proposing leave and leave forms seemed to be quite a quandary in this organization. First, the issue of the requestor who does not sign his leave request form. Second, there was no copy of the approved form for the requestor. Now this person was acting as if she was doing me a favor for correcting the problem in which she created. It was she who did not read the email that was sent to her clearly stating the changes that needed to be made.

The seventh day of October quickly raced in and once again a new batch of personnel was introduced into the work environment. There were six new personnel added this time around. Just days before the influx of new personnel, Jose and Tex were transferred to new assignments. Jose went to Tagi which was a small station slated to soon close down. Perry was slated for Balaud, which was a huge Air Base with excellent facilities. However, he would not leave for another week. Tex was requested by name for an assignment in Kuwait thanks to his friend Jeffery who put in a *good word* for him. Lenard had the good fortune of remaining with me in good old Baghdad. Three days after Tex arrived in Kuwait, he sent me an e-mail stating, *"I don't think I am going to like it here."*

Alfred announced there was a permanent change assignment that required two FSRs. If there were no volunteers there would again be persons *volun-told*. Stan did not have the best of relations with his DoD counterparts and quickly volunteered. The next day Bailey volunteered for that same assignment. After Stan heard that Bailey also volunteered for the same assignment he tried to get out of it, but his name had already been sent forward. Stan then tried to persuade me to volunteer for it in hopes that I would be selected instead of Bailey, but I was not interested in going to a small FOB that had been said to constantly have dust storms. They both left about a week later and were never heard from again.

One of the new arrivals was a 45 year old black male named Nathan. He was light skinned and wore his hair in braids. Nathan was a retired US Army Sergeant 1st Class. He was new to the WarWizard Program. After we formally introduced ourselves Nathan explained that he was supposed to be assigned to Kuwait but was assigned to Baghdad instead. Nathan was really bummed out about this because his wife and six year old daughter had settled in Kuwait based on his projected assignment to Kuwait. Nathan stressed that he was afraid to press the issue of alternate placement because he was a new hire early in his probationary period with the company.

When I heard this I quickly steered him in a practical direction saying, "Just days ago, there were three guys transferred from here to Kuwait. So if you want to be transferred, talk to Rusty about it. Your contract says Kuwait and right now they are sending personnel there." "You don't think it would be a problem at all since I am so new?" Nathan questioned. "The time is now, so if I were you I would talk to Rusty. Just tell him the same thing that you told me about your wife and child" I assured him. Well, our brief casual conversation was all it took because after Nathan talked to Rusty, two days later Nathan was on the first thing smoking to Kuwait. Nathan had thanked me for my insight and offered to me if there was ever anything that he could do for me to let him know. I told him that I was just glad that I could help.

I had grand anticipation of the Monday Night Football match-up between New Orleans and the Giants because only one team would remain undefeated at the end of this NFL *Game of the Week*. Since Perry was leaving tomorrow for his permanent duty assignment we decided to spend his last night in Baghdad hanging out and watching the game. We chose to watch the game in the bunkhouse since it was equipped with a fifty two inch TV. New Orleans offense got off to a fast start with several touchdowns. That night the Saints had its way with the New York Giants and at the games conclusion the Saints remained undefeated with a perfect 5-0 season; final score New Orleans 48, Giants 27. The only occurrence that would have made this win better was if Jojo could have been there to share this glorious moment with us. Fortunately for Jojo, he had already been transferred to Kuwait on TDY to assist with their maintenance.

CHAPTER 19

KUWAIT REVISITED

October was drawing to an end with it already being the 25th day. Things were slow at work that day and about ten of us were gathered in the break room when Alfred walked in. He informed me of my selection by the Headquarters Staff to be a member of a maintenance installation project in Kuwait. I inquired as to how long the duty assignment would last? Alfred's response was he could not give me an exact timeframe but speculated it to be somewhere between one and six months. Alfred also informed that I must be in place in Kuwait by 29 October. That initially did not sound too bad but as with most things it comes with a catch; "By the way Richard is also going" uttered Alfred as he exited the room as if he was aware that he had covered pretty good news with bad. I was somewhat excited on one hand but had a negative outlook on the other. After all getting out of a war zone was a very good thing. My negative outlook stemmed from all the negative press I had heard about the Kuwait WarWizard Facility. I had no idea what was true and what was not. Also being that Richard was included in the deal made the venture not as appealing; I just can't seem to break free of that guy.

Another positive was the fact that it seemed to be a good thing when you got the chance to leave Baghdad and see other places in the region. Besides, I had I just recently volunteered to be a member of a traveling maintenance team; I just didn't voice my wishes early enough and by the time I did the assignment was no longer an option. So going TDY for the most part sounded pretty good because it seemed to help the time go by quicker.

The projected time to leave squashed my plans to attend a Halloween costume party where the admission was your costume; no costume no admission. It sounded like a fun way to break up the monotony. I had already gathered the makings of a Zorro Costume (a black hat, shirt and slacks, plus black material for a cape and an old broken antenna for my sword). The next day, to add insult to injury, I learned that Jameson was added to the Kuwait TDY. All is well though, because when duty calls you have to answer; after all, that's what for.

What ever happened to Richard not having to go on any TDYs that he did not desire to attend? That matter would soon be put to the test. After the TDY was made known to him he immediately began his efforts to get out of it. After all, his family had Deep South Connections with Edmond Gibson the Program Manager; at least that was the line he had given me weeks earlier. But it soon looked as if the Mississippi edge that Richard spoke of did not exist and could be classified as another Richard myth. This conclusion was made because after all the dust settled with his objections, Richard was on the hook to go to Kuwait just as the rest of us.

When I arrived to work the next morning there were still three eighteen wheel transport vehicles in the bay from the previous night. Lenard was a quick study and had really learned his craft well with these types of installations, thus he headed up the project to complete those installs. Seeka (a new guy from Samoa who worked at my friend Bert's FOB) was also a key factor that made Lenard's project run well. I had just finished working with Lenard on his install project when Marshall came over and asked me to help Richard and Jameson's' team get their truck finished because they were behind schedule. I wanted no part of this because weeks earlier I had specifically requested and had permission granted by Malcolm not to work with Richard because we did not work well together. But since I was asked by Marshall to help I did not resist the request.

As I began working on the vehicle it was clear why that team was behind. Other team members were working but Richard and Jameson were talking to their minister. This was the same guy who performed Richard's baptism weeks earlier; I will refer to him as the minister. I never let the minister or anyone else know this, but I did not give much credence to him as a Minister. After all he is the one that appointed Jameson as Bishop. I continued to work, but with skeptical ears I listened to their minister as he spoke to Richard and Jameson, "God knew what he was doing when he first sent Richard to Kuwait; God knew that Richard needed a spiritual advisor so he grabbed the Bishop and sent him to watch over brother Richard. So I know Richard will be alright, but the real question is how is brother Kingston going to serve God when he gets to Kuwait." "Hey, you don't have to worry about me because me and God are okay" I stated. "Oh so you are self righteous," said the Minister. "The Bible speaks against those who are self righteous" Jameson added. I paused for a second and directed my gaze at Jameson and the minister then commented, "You two got all of that out of what I said? There was complete silence between them both. "Well if you did good for you." I then turned my back to them and continued on with my work. The minister left a few minutes later, although his exit served as no clue to get back to work because Richard and Jameson screwed around and joked on top of the truck for another 15 minutes or so.

I was working at ground level, while Richard and Jameson were on the top of the vehicle still joking. Keep in mind the Assistant Floor Lead Marshall asked me to assist those two because they were behind with their install. As I continued to work, out of the blue Richard uttered "Kingston, you have a long ass head." My response was immediate, "Richard, everything about me is longer than you, my arms, legs and my brain and everything else are all longer than yours." The arms comment may have sent him over the edge because Richard has slightly abnormally short arms. Darrell Ryan had often teased Richard by saying he had arms like a T-Rex. Then Richard angrily uttered more irreverent comments, "You think you are so smart, then tell me why I make more money than you if you are so smart?" "How do you know you make more money than I do?" I questioned. "Because I have been in the program much longer than you," Richard responded. I responded at him saying, "You enter this program based on your skill set. Must I inform you that I have a Masters degree, retired Air Force and years of electronics and maintenance experience; that alone trumps me ahead of you. So I'm sure I entered the program making a larger salary than when you entered at the base level and probably still earn more than you earn right now. Another thing, since you brought up the *smart* issue, I'll be glad to debate you and your idiot spiritual advisor on the Bible any time since you guys saw fit to tell me where I am on the spiritual subject." "Why would I debate you on a subject that I know you are strong in?" Richard questioned. "I don't think it is a matter of I know so much on the subject, it's that you don't know anything on the subject" I replied. "You have just one more time to question my intelligence," Richard shouted to me as he angrily pointed his finger down at me. "I can say whatever I want whenever I want," I sternly responded as I pointed my finger back at him. By now it was obvious that Richard was highly frustrated, he then said "You wait right there; Today we both will be sent home." Richard jumped off the truck like he was a crazed dog, eager to attack. He ran up to me in a manner as to solicit some sort of a reactive move, but I did not flinch. He just stood there breathing very intensely. I then raised my arm and said, "someone better come and get this guy; SOON!" Malcolm rushed over to where we were standing and said "Kingston its lunch time," and released me for lunch.

As company policy states, anyone who engages in any type of physical altercation could result in both parties being terminated. I will admit that things escalated fairly quickly and could have gotten out of control, but one thing was for certain; I was not going to let someone like Richard be the cause of ending my livelihood prematurely. I would consider Richard a unique character. In the words of Forrest Gump, Richard is "like a box of chocolate, you never know what you going to get." But Richard was now a bigger joke than he was when he had arrived for work that day because everyone already knew he was all talk on most things that mattered, but now even on thug type issues.

It was just days before my proposed departure to Kuwait and I had not been given any additional information, flight itinerary or anything. Therefore, that morning when I arrived to work, I questioned Alfred for some firm guidance about the trip. Alfred informed me that he had not been given any more details about the TDY other than a program name; *"Dash Eight,"* which he had already revealed to me. Alfred offered me a suggestion saying, "Maybe you can ask Richard what does he knows, after all he is married to Bernice and she works at the Headquarters Office." "Alfred, I do not want to place any faith on any trip or anything else to rest on the words of Richard" I replied. "Well Kingston that's just one more thing that you and I have in common because I would not want to place any faith on anything dealing with that guy as well. Would you believe that he asked me for two days off to pack?" Alfred continued. "Really, how did you respond to his question?" I questioned. "I told him if Rusty says its okay then I will go with it. Then Richard said that was okay" Alfred replied laughing. Continuing to laugh as he walked off, Alfred promised he would contact Bernice about the program details.

On the day before our Kuwait departure there was a company sponsored picnic that started at 4:00pm. I was released at noon to pack for Kuwait. I estimated that I would try to attend the picnic later after packing and tying up any loose ends. I was unable to finish packing by the time it reached 4:30pm so I decided to take a break to attend the company event. There was plenty of activity going on; most of the personnel were outside in shaded areas while others were inside along with the food. I spend enough time in the heat I thought, so I made my way inside.

The crowd dissipated to around fifteen in a matter of about thirty minutes. "Anyone for spades" Kimberley shouted? Out of the stragglers left in the room, Richard's wife and I were the first to agree to play. "We need one more," Kimberley voiced. "I guess I will go ahead and play," voiced Vanessa (the Contract Monitor who attempted to make a case about the football toss weeks earlier). By the time the game was underway there were only four people left in the room; three females and myself. Vanessa and I teamed up and we got off to a good winning start. About that time Richard entered the room. We made eye contact for a second then he had a seat on the sofa. Based on the expression on his face he appeared as though he did not appreciate me sitting at a table in the company of all females with one of them being his wife.

However, I was content, it was just an innocent card game. No harm, no foul as Vanessa and I continued to demolish our opponents. During the course of the game Bernice informed me that she would be taking the Kuwait TDY Team to the air terminal the next morning with a 5:00am show time. The conversation migrated to the Halloween costume party I would be missing due to my flight in the morning. Kimberley mentioned that she was going to the party in a Saint / Sin costume. I asked her to explain her contradictive costume. She then informed, "It is half angel and half devil." "Now I am really going to hate missing that costume party" I voiced. The game ended with Vanessa and me winning all but one game. It was a fun way to end the evening before going off into the unknown waiting for me in Kuwait.

When I got to my room that evening, I resumed packing. I packed as much as I could because I had no idea when I would be back. By the time I completed the whole packing process it was 3:30am, which gave me only about forty minutes to sleep. Jameson and I were in place on time for Bernice and Richard to pick us up for transport to the Air Terminal. They ran about fifteen minutes late, but we were soon at the terminal in plenty of time to make our flight departure. Bernice didn't just drop us off at the air terminal she and Richard spent that time together until we boarded the plane.

Just as it was when I entered the country, the plane that we exited the country in was a good old US Air Force C-130. The main difference this time was no combat landing was necessary because we were landing in peaceful air space. The projected flight time was approximately one hour. The flight passengers consisted of four civilians and twenty-five soldiers. Of the twenty-five soldiers, three were Caucasian females; one lieutenant and two enlisted. One of the enlisted female soldiers and the female officer took many pictures of the flight (inside and outside aircraft) and their fellow snoozing soldiers. I was requested and took a picture of the three of them. These same females offered to carry some of my huge luggage collection once we landed at

Ali Al Saleem, Air Base Kuwait. I thanked them for their offer and I had a lot of luggage, but it just didn't feel right having them carry my luggage so I refused. Next the three of us signed in the base with our CAC Cards and then turned in our bullet proof vest and helmets into the holding facility. This exchange was done to avoid the continued carry of the extra equipment weight. Finally, it was time to get shelter through billeting in a nice air conditioned tent for the next four to five hours until the Kuwait WarWizard personnel picked us up. The last that I had heard from Tex was an Email that stated he did not think he was going to like it here. So it was very much a question mark as to what I was to face as we now sustained Operation Iraqi Freedom's new level of Knife Wounds In Kuwait.

GLOSSARY

- **Airman First Class (A1C)** Air force enlisted rank of E-3.

- **APOD** Aerial Port of Debarkation: the military side (USAF) of Kuwait's International Airport.

- **Ali Al Salem Air Base** is a military airbase positioned in Kuwait approximately twenty miles from the Iraqi border. The airfield is owned by the Government of Kuwait. The airfield served as a staging point during Operation Southern Watch, Operation Enduring Freedom, Operation Iraqi Freedom and Operation New Dawn.

- **Arifjan** Camp Arifjan is a United States Army installation located in the State of Kuwait which accommodates elements of the U.S. Air Force, U.S. Navy, U.S. Marine Corps and U.S. Coast Guard. Camp Arifjan is located south of Kuwait City and Al Salem Airbase.

- **Baghdad** is the capital of the Republic of Iraq and resides on the Tigris River.

- **Balad Air Base** was located approximately fifty miles north of Baghdad in the Salah al Din Governorate Iraq. It was located within the borders of the so-called Sunni Triangle; however, Balad is a primarily Shiite town of approximately 100,000. The Air Base was also known as Logistics Support Area Anaconda, or simply LSA Anaconda. It was formerly the largest Iraqi Air Force base during the Saddam Hussein era.

- **Blue Force Tracking** is a United States military phrase used to denote a GPS-enabled system that provides military commanders and forces with location information about friendly and hostile military forces.

- **Burqa** is an enfolding outer item of clothing worn by women in some Islamic traditions to cover their bodies when in public; burqa are usually black.

- **Campaigns Designation** Official name of a military operation or war (ie., Operation Iraqi Freedom).

- **Civil ID** is a requirement for any expatriate or resident regardless of the type of residence he owns. It is a card that includes the card number, full name, date, birth, local address, blood type, sponsors name, and type of visa and card expiration. The life cycle of a Civil ID is two years.

- **Common Access Card (CAC Card)** Basic identification card necessary to eat in the DEFAC, fly on military flights and many other identification purposes..

- **Criminal Investigation Division (CID)** United States Army Criminal Investigation Command (USACIDC, regularly abbreviated as only CID). CID investigates felony crimes and severe violations of military law within the United States Army.

- **Contingency Operating Base (COB)** any secured military position, normally a military base, used to support tactical operations.

- **Department of Defense (DoD)**

- **Defense Information Network (DSN)** Sensitive but Unclassified voice service.

- **Dining Facility (DEFAC)** the primary eating setting for both civilian and military in Iraq.

- **Explosive Ordnance Disposal (EOD)** the removal and disposal of bombs.

- **Flak vest** is a form of protective clothing designed to provide protection from Flak shells, shrapnel and other indirect, low-velocity projectiles.

- **Fly-Aways** Helicopter flights that transport contractors to other locations to assist.

- **Forward Operating Base (FOB)** any secured forward military position, normally a military base, used to support tactical operations.

- **Forty Site Install Office** Located on the Forty Site which is twelve miles north of Camp Arifjan. The mission of the install office is to perform installs on systems in convoy vehicles. The install office is subordinate to the Operations Manager located at Arifjan.

- **Forty Site Yard Team** Group located on Forty Site's yard. Assembled for high volume of De-installs. Yard Team Lead was Subordinate to Forty Site Lead, who is located in the Install Office of the Forty Site.

- **FSR Field Service Representatives** Performed maintenance on jammer systems

- **Ganymede Dynamics** Fictitious name for a real company that was the largest private sector entity on the Forty Site.

- **Guerrilla warfare** is a form of irregular warfare in which a small group of combatants use military tactics such as ambushes, raids, and sabotage. The element of surprise and the ability to move quickly is the strength of this type of army verses a larger and less-mobile traditional military.

- **Helo Pad** Helicopter landing and launch location.

- **Halliburton** is one of the world's largest oilfield services companies with operations in more than seventy countries. It has hundreds of subsidiaries, affiliates, branches, brands, and divisions worldwide and employs over 60,000 people. Halliburton's former subsidiary, KBR, is a major construction company of refineries, oil fields, pipelines, and chemical plants. In November 2002, KBR was tasked to plan oil well firefighting in Iraq, and in February 2003 was issued a contract to conduct the work. Critics contend that it was a no-bid contract, awarded due to Dick Cheney's position as Vice President. The company has dual headquarters located in Houston and in Dubai.

- **Henry Kissinger** U.S.diplomat of Jewish heritage but German born. Nobel Prize Winner and served as the National Security Advisor, also served as the Secretary of State in bothe the Nixon and Ford administrations.

- **Improvised Explosive Device (IED)**, also referred to as a **roadside bomb**, is a homemade bomb forged and deployed in ways other than in typical military action. It may be constructed of conventional military explosives, such as an artillery round, attached to a detonating mechanism.

- **KBR, Inc.** (formerly **Kellogg Brown & Root**) is an American leading global engineering, construction and private military, government services and civil infrastructure contracting company. In addition, services company supporting the energy, petrochemicals. Formerly a subsidiary of Halliburton, headquartered in Houston, Texas.

- **Lot 58** Location on Arifjan that contains military vehicles; occupational location of the WarWizard De-Install Team located on Arifjan.

- **Lock-&-Key Machinery** Fabricated name for a real company that supported the war effort in Iraq.

- **High Mobility Multipurpose Wheeled Vehicle (Humvee)**, is a four-wheel drive military automobile that replaced the jeep.

- **Operations Manager** Highest ranking person in the WarWizard Program in Kuwait. Subordinate to Chief of Maintenance located in Baghdad.

- **Max-Pro Basic** An MRAP with the minimum armor.

- **Mine-Resistant Ambush Protected (MRAP)** vehicles are a family of armored fighting vehicles used by the United States Armed Forces. The purpose of the design is surviving Improvised Explosive Device (IED) ambushes and attacks.

- **Navy E7** Enlisted Grade 7, Chief Petty Officer

- **Navy 05** Officer Rank of Commander

- **Non-commissioned officer (NCO)** a sub officer in the military who has not been given a commission.

- **Red herring** a clue which is misleading; sometimes intentionally and other times unintentionally or simply something that distracts from the actual issue.

- **RG-33** a mine-resistant light armored military vehicle (truck).

- **Shiite** a member of a division of Islam that regards Ali and his descendants as the rightful successors to Muhammad and discards the first three caliphs.

- **Southwest Asia (SWA)**

- **Sunnis** The branch of Islam that approves the first four caliphs as legitimate successors of Muhammad.

- **The Palace (The Al-Faw Palace)**,one of Sadam's former Palaces which served as the headquarters for the Multi-National Corps for Iraq and later United States Forces. It was positioned on Camp Victory.

- **Taji** was a FOB situated approximately twenty miles northwest of downtown Baghdad near the city of Tagi. It was one of the U.S. Army's operational hubs ground tactical and unmanned aerial vehicles. Taji was turned over to the Iraqi Government in 2010.

- **Tallil Air Base** was a military airbase situated near Nasiriyah, Iraq. The installation originally served as an Iraqi Air Force Base named "Imam Owen Air Base." It is commonly known as "Camp Adder" by the U.S. Army; the name "Owen Air Base" is used chiefly by the U.S. Air Force. The ancient Babylonian city of Ur, recognized as the birthplace of Abraham, is located within the security perimeter for Owen Base, and its ancient Ziggurat is also visible from the base. The city of Nasiriyah is located on the Euphrates River. Until December 2011, the base was used by United States Armed Forces.

- **Temporary Duty Assignment (TDY)**

- **Third World Country National (TCN)**

- **Triton Telecommunication Inc. (TTI)** Fictitious name for a real company that was the largest private sector entity on the Forty Site.

- **T-Wall** was a concrete upside down T that was about 2 feet thick and about 12 feet tall. The T was inversed so that the wall would stand. The T-Walls function was to serve as a protective perimeter that prevented a possible enemy from having the capability to do drive-by's.

- The **Wire** the premier or boundary of the base (COB or FOB).

- **Space A** Vacant Seats granted on military aircraft on a first come first serve basis. Managed with a sign up roster. Emergency have priority over casual passengers.

- **Sea Port of Debarkation (SPOD)** Naval and Cost Guard Shipping Port Located East of Camp Arifjan.

- **SRA (Senior Airman)** Air Force Enlisted grade of E-4.

- The **United Arab Emirates** (**UAE**) is a constitutional association of seven emirates that include Abu Dhabi, Ajman, Fujairah, Umm al-Qaiwain, Ras al-Khaimah, Sharjah, and Dubai. On December 2, 1971 the alliance was officially recognized.

- Camp **Victory** was located in Baghdad. This was the major component of the Victory Base Complex (VBC) which occupied the area surrounding the Baghdad International Airport (BIAP). Other camps that made up the Victory Base Complex incorporated Camp Liberty (formerly known as Camp Victory North), Camp Striker, and Camp Slayer. The Tigris River ran right through the complex. Under agreement with the Iraqi Government in 2008, Camp Victory was handed over by the United States to the Iraqis in December 2011.

- **Visa 14** A Temporary Visa with a timeline of 90 days. Until a Residency Visa is obtained, the holder must leave the country ever 90 days.

- **Visa 18** A Residency Visa with a timeline of one year. If holder is out of the country for more that 6 months the Visa is canceled.

Epilogue

JOHN CHANCEY KINGSTON is a modern approach author with an older time prospective of good and evil coupled with his unique methods of presenting his ideals.

In coming months Part II of this trilogy entitled **Knife Wounds From Desert Sand: Darker Wars Revealed Behind Contractor Support in Kuwait & Iraq** will be released.

In his first book **If You Want An Enemy, Try to Help Someone** Chancey presented an even more sinister side of human nature.

He revealed the recipient of help as someone who will sometimes blame the helper if things turn out bad or, possibly even worse, make them and enemy.

Made in the USA
Monee, IL
04 November 2026

67377168R00095